Biotechnology Education, Outreach and Extension

Biotechnology Education, Outreach and Extension

P. R. Gilbert

ANMOL PUBLICATIONS PVT. LTD.
NEW DELHI - 110 002 (INDIA)

ANMOL PUBLICATIONS PVT. LTD.
H.O.: 4374/4B, Ansari Road, Darya Ganj
New Delhi-110 002 (India)
Ph.: 23278000, 23261597
B.O.: 1015, Ist Main Road, BSK IIIrd Stage
IIIrd Phase, IIIrd Block
Bangalore - 560 085 (Karnataka)
Ph: 080-41723429, Tel/Fax: 080-2672 3604
Visit us at: www.anmolpublications.com

Biotechnology Education, Outreach and Extension

First Published, 2008

ISBN 978-81-261-3277-5

PRINTED IN INDIA

Printed at Mehra Offset Press, Delhi.

Contents

Preface

This book entitled "Biotechnology Education, Outreach and Extension" is designed to acquaint readers with the modern *biotechnology* education, outreach and extension at the global level. Biotechnology is the use of biological systems, i.e. living things, to make or change products. It has been used for centuries in traditional activities like baking bread and making cheese. Traditional biotechnology involved for example, developing a new wheat variety with early ripening characteristics by cross-breeding different types of wheat until the desired characteristics and only those characteristics were present in the new wheat variety. The terms 'gene technology', 'genetic engineering' and 'genetic modification' mean the same thing, and refer to one type of modern biotechnology. Since the 1940s scientists have known that DNA — deoxyribonucleic acid — in the cells of all living things is like a blueprint that is passed from one generation to another. Genes are made of DNA. They contain coded instructions for proteins, which give living things their particular characteristics like hair and eye colour. All living organisms use DNA as the genetic code. Gene technology includes a range of techniques to copy genes and modify them so that they will work in a new host either to produce or reduce a product. It also involves transfer of the reconstructed gene to a new host. Using gene technology, scientists aim to introduce, enhance or delete particular characteristics of a living thing, depending on whether they are considered desirable or undesirable. As you go further into this website, you can read more about gene technology — its history, how it's done, its possible uses, the regulatory processes in place for gene technology, the debate surrounding gene technology and how this technology might affect you. Gene technology is a term that refers to a whole range of techniques for genetic analysis that depend on the direct manipulation of DNA, the material substance of heredity. The use of gene technology in biomedicine, agriculture, food production and processing is an issue that has evoked strong public interest and concern. The debate has focused on the use of genetically modified organisms (GMOs) in food, their safety and potential impact on the environment. The public better accepts medical and industrial uses of gene technology. The majority of Australians are eager for more quality information about the technology and its applications. Most people think of gene technology as adding genes from other life forms to plants and animals. While this is certainly part of it, the technology is broader than that, and has many more uses. Gene technology is developing at a rapid rate and it will continue to revolutionise basic biological research and development. It provides the potential to improve our health, create a safer and more secure food supply, generate greater prosperity and attain a more sustainable environment. Gene technology is already providing new ways of preventing, treating and curing human and animal diseases; it is helping farmers improve agricultural production with less impact on the environment; and in the near future, it will allow better food products to be available to consumers at reduced cost.

The technology has given us vital new products like human insulin for diabetes, interferon and other drugs for treating certain cancers, and vaccines against diseases like hepatitis B. Millions of human lives are protected by gene technology every day. Genes are made from a chemical called DNA (deoxyribonucleic acid) that is common to all forms of life on Earth. Genes carry information for making all the proteins required by all organisms. These proteins determine, among other things, how the organism looks, how well its body uses food or fights infection, and sometimes even how it behaves. DNA is made up of four similar chemicals (called bases and abbreviated A,T, C, and G) that are repeated millions or billions of times throughout a genome. A genome is the entire DNA in an organism, including its genes. The human genome, for example, has 3 billion pairs of bases. The particular order of As, Ts, Cs, and Gs is extremely important. The order is called DNA sequence and variations in sequence underlie all of life's genetic diversity. DNA sequence is the database determining how the cell is to make proteins and what functions the proteins can perform. Because all organisms are related through similarities in DNA sequences, insights gained from non-human genomes often lead to new knowledge about human biology, and insights from the genes of a bacterium, worm or bee might contribute to new solutions for the wheat farmer. Gene technology consists of tools and techniques that scientists can use to study, identify or modify the genes of living organisms. Scientists can use gene technology to locate and analyse single genes in a chain of many thousands. They can make gene probes that recognise DNA sequences associated with genetic diseases. The diagnostic tests involve analysing minute tissue samples collected from adults, or taken from embryos. This information can assist parents and doctors to reduce the incidence of crippling genetic defects, and treat other disorders much better. No two individuals have the same genetic makeup, a fact that is now being used by forensic scientists to identify individuals from cell samples left at a crime site. "Forensic DNA fingerprinting" involves taking a blood or tissue sample from a person and using one of a number of gene technologies to generate patterns of DNA fragments separated by size. Those patterns look something like a long bar code and are dependent on the exact sequences in the sample. A match between the genetic fingerprint from the crime sample and the suspect has been admissible evidence for a number of years. For centuries plants and animals have been selectively cross-bred for desirable traits such as size, enhanced quality and pest resistance. Scientists can now use DNA information to identify desirable traits in organisms.

The antibiotics are used only in the laboratory and are of no medical significance. The antibiotic resistance genes employed are already widely distributed through the natural environment, such that their presence in the plant is of no environmental consequence. Nevertheless, because of public concerns about antibiotic-resistance genes, researchers have developed ways to remove these gene markers from plants before they are released commercially. New types of marker genes have also been developed, such as those permitting growth in the laboratory on unusual sugars. A gene that is producing undesirable characteristics in an organism can be turned down or switched off. One way this can be achieved is by inserting a second copy of the gene, or a fragment of the gene, back to front. Other ways include RNAi and "gene shears", which cause the gene message molecule to destroy itself. This cancels the effect of the undesirable gene. Many human and animal diseases have now been traced to their host being born with faulty genes that produce defective proteins. These diseases include sickle cell anaemia, cystic fibrosis, the blood clotting disorder haemophilia and Down's syndrome. Modern medicine and surgery have made great strides in treating patients with genetic diseases. Scientists are now exploring the use of gene therapy to attack these diseases at their source — by correcting faulty genes. Gene therapy holds great promise for treating disease by replacing or changing a very small part of the overall genetic program of carefully selected cells, perhaps permanently, producing a cure. It aims to restore the healthy function of cells by replacing or correcting the defective gene. Gene therapy can be used to replace an abnormal gene with a normal one, to insert a missing gene, to switch off rogue genes that may cause cancers and to stop viruses multiplying within cells. The modified cells and genes are not passed onto children. Certain technical obstacles are yet to be overcome if gene therapy is to deliver the expected benefits to treat human disease. Among the obstacles is the lack of a delivery system, or vector, that can safely and efficiently shuttle beneficial genes into the cells of patients — and ensure they work. Even the most advanced cell therapy techniques are still at the experimental stage in clinical trials. Further research is needed to develop safe, reliable gene therapy techniques.

This book consists of—General Course Description of a Biotechnology Program; Model Educational Programs, Causes and Sample Lessons in Biotechnology; Select Biotech Information and Capacity Building; and Sample Academic Programs in Biotechnology etc. besides a large body of list of acronyms, glossary of relevant terms, extensive bibliography, list of websites and links, for further referencing and research.

—Editor

List of Acronyms

AAAI	—	American Association forArtificial Intelligence
AAAS	—	American Association for the Advancement of Science
AACC	—	American Association for Clinical Chemistry
AACR	—	American Association for Cancer Research
AAI	—	American Association of Immunologists
AANP	—	American Association of Neuropathologists
AAP	—	Association of American Physicians
ABA	—	Australian Biotechnol. Assoc.
ABI	—	Applied Biosystems, Inc.
ABIC	—	Agricultural Biotechnology Intl. Conf.
ABRF	—	Association of Biomolecular Resource Facilities
ACMG	—	American College of Medical Genet.
ACS	—	American Chemical Society
ADA	—	Americans with Disabilities Act
AEC	—	Atomic Energy Commission
AES	—	American Electrophoresis Society
AFMR	—	American Federation of Medical Research

AFIP/ARP	—	Armed Forces Inst. of Pathol./Am. Registry of Pathol.
AGSG	—	Alliance of Genetic Support Groups (now Genetic Alliance)
AGT	—	Association of Genetic Technologists
AHA	—	American Heart Association
AIBS	—	Am. Inst. of Biol. Soc.
AMIA	—	American Medical Informatics Association
ANGIS	—	Australian National Genomic Info. Service
ANL	—	Argonne National Laboratory
API	—	Application Programming Interface
ARP	—	American Registry of Pathology
ASB	—	American Society for Biotechnology
ASBMB	—	American Society for Biochemistry & Molecular Biology
ASCB	—	American Society for Cell Biology
ASCI	—	American Society for Clinical Investigation
ASHG	—	American Society for Human Genetics
ASIP	—	American Society for Investigative Pathologists
ASIS	—	American Society for Information Science
ASLME	—	American Society of Law, Medicine, and Ethics
ASM	—	American Society for Microbiology
ASPET	—	American Society for Pharmacology and Experimental Therapeutics

ATCC	—	American Type Culture Collection
ATP	—	Advanced Technology Program
AVS	—	American Vacuum Society
AWCH	—	Adelaide Women and Children's Hospital
BAC	—	Bacterial Artificial Chromosome
BACR	—	British Association for Cancer Research
BCM	—	Baylor College of Medicine
BDG	—	Batten's disease gene
BER	—	Biological and Environmental Research
BIO	—	Biotechnology Industry Organization
BNL	—	Brookhaven National Laboratory
BSCS	—	Biological Sciences Curriculum Study
BS/SCF	—	Biological Sequence/Structure Computational Facility
BTCI	—	BioPharmaceutical Technology Center Institute
BTP	—	Biotechnology Training Programs
CAE	—	capillary array electrophoresis
CaSSS	—	California Separation Science Society
CATCMB	—	Center for Advanced Training in Cell and Molecular Biology
CCM	—	Chromosome Coordinating Meeting
CDC	—	Centers for Disease Control
cDNA	—	Complementary DNA
CE	—	Capillary Electrophoresis
CEPH	—	Centre d'Etude du Polymorphisme Humain

CF	—	Cystic Fibrosis
CFF	—	Cystic Fibrosis Foundation
CHH	—	Cartilage-hair Hypoplasia
CHI	—	Cambridge Healthtech Inst.
CHOP	—	Children's Hospital of Philadelphia
CIMB	—	Center for International Meeting on Biology
CIOMS	—	Council for International Organizations of Medical Sciences
CIRB	—	Colorado Institute for Research in Biotechnology
CLMA	—	Clinical Laboratory Management Association
cM	—	Centimorgan
CMT	—	Charcot-Marie-Tooth
CONTIG	—	Consortium of Teachers in Genetics
CORN	—	Council of Regional Networks for Genetic Services
CRADA	—	Cooperative Research and Development Agreement
CSHL	—	Cold Spring Harbor Laboratory
CTG	—	Trinucleotide Repeat
DHHS	—	Dept. of Health and Human Services
DIMACS	—	Center for Discrete Math. & Theoretical Comp. Sci.
DM	—	Myotonic Dystrophy
DMD	—	Duchenne Muscular Dystrophy
DNA	—	Deoxyribonucleic acid
DOE	—	Department of Energy
ECB	—	European Congress on Biotechnology

EDS	—	Electronic Data Submission
EEOC	—	Equal Employment Opportunity Commission
EL.B.A.	—	ELectronics and Biotechnology Advanced
EMBL	—	European Mol. Biol. Lab.
EMBO	—	European Molecular Biology Organisation
EMG	—	Encyclopedia of the Mouse Genome
EMS	—	Environmental Mutagen Society
EORTC	—	European Organization for Research and Treatment of Cancer
ERDA	—	Energy Research and Development Administration
ERI	—	Eleanor Roosevelt Institute
ES	—	Embryonic Stem
ESF	—	European Science Foundation
ESHG	—	European Society of Human Genetics
ESI	—	Electrospray Ionization
EST	—	Expressed Sequence Tag
EUCIB	—	European Collaborative Interspecific Backcross
EURESCO	—	European Research Conferences
FAP	—	Familial Adenomatous Polyposis
FASEB	—	Federation of American Societies for Experimental Biology
FCM	—	Flow Cytometry
FEBS	—	Fed. of Eur. Biochem. Soc.
FMF	—	Familial Mediterranean Fever
FRAXA	—	Fragile X locus

FSHD	—	Facioscapulohumeral Muscular Dystrophy
FTICR	—	Fourier Transform Ioncyclotron Resonance
FVEA	—	Fundacion Valenciana de Estudios Avanzados
GAS	—	Genome Automation System
GBASE	—	Genome Database of the Mouse
GBR	—	Global Business Research
GDB	—	Genome Database
GDB/OMIM	—	Genome Database/Online Mendelian Inheritance in Man
GESTEC	—	Genome Science and Technology Center
GIST	—	Genome Informatics System of Transputers
GLaRGG	—	Great Lakes Regional Genetics Group
GMCRF	—	General Motors Cancer Res. Foundation
GMD	—	Genomic Map Design
GPI	—	Genetics and Public Issues Program
GRAIL	—	Gene Recognition and Analysis Internet Link
GRC	—	Gordon Res. Conf.
GSA	—	Genetics Societies of America
GSDB	—	Genome Sequence Data Base
HAEC	—	Human Artificial Episomal Chromosome
HELSRD	—	Health Effects and Life Science Research Division

HERAC	—	Health and Environmental Research Advisory Committee
HGCC	—	Human Genome Coordinating Committee
HGM	—	Human Genome Meeting
HGMIS	—	Human Genome Management Information System
HGP	—	Human Genome Project
HHMI	—	Howard Hughes Medical Institute
HICSS	—	Hawaii Intl. Conf. on Systems Sci.
HLA	—	Human Leukocyte Antigen
HMDP	—	Homology Database
HNPCC	—	Hereditary Nonpolyposis Colorectal Cancer
HSA	—	Human Serum Albumin
HUGO	—	Human Genome Organisation
IARC	—	International Agency for Research on Cancer
IBC	—	International Business Communications
IBEX	—	International Biotechnology EXpo
IBF	—	International Business Forum
IBI	—	Institute for Biotechnology Information
ICES	—	International Council of Electrophoresis Societies
ICGEB	—	International Centre for Genetic Engineering and Biotechnology
ICHG	—	International Congress of Human Genetics

ICPEMC	—	International Commission on Protection Against Environmental Mutagens and Carcinogens
ICRF	—	Imperial Cancer Research Fund
IEEE	—	Institute of Electrical and Electronics Engineers
IG	—	IntelliGenetics
IGES	—	International Genetics Epidemiology Societies
IJCAI	—	International Joint Conference on Artificial Intelligence
IMA	—	Institute for Mathematics and its Applications
IMACS	—	International Association for Mathematics and Computers In Simulation
IMAGE	—	Integrated Molecular Analysis of Gene Expression
IMGS	—	Intl. Mammalian Genome Society
INRIA	—	French Natl. Inst. for Research in Computer Science and Control
IOM	—	Institute of Medicine
IOR	—	Institute of Religion
ISAG	—	Intl. Society for Animal Genetics
ISMB	—	Intelligent Systems for Molecular Biology
ISONG	—	Intl. Soc. of Nurses in Genet.
ISQL	—	Interactive Standard Query Language
ISTR	—	Institute for Science Training and Research

IUBMB	—	Intl. Union of Biochemistry and Molecular Biology
IU	—	Indiana University
IVD	—	*in vitro* Diagnostic
JGI	—	Joint Genome Institute
LANL	—	Los Alamos National Laboratory
LBNL	—	Lawrence Berkeley National Laboratory
LLNL	—	Lawrence Livermore National Laboratory
LTI	—	Life Technologies, Inc.
MALDI	—	matrix-assisted laser desorption ionization
MBC	—	Massachusetts Biotechnology Council
MBL	—	Marine Biological Laboratory
MCD	—	Mouse Cytogenetic Database
MDA	—	Muscular Dystrophy Assoc.
MEMS	—	Microelectromechanical Systems
MGC	—	Mouse Genome Conference
MGD	—	Mouse Genome Database
MGI	—	Microbial Genome Initiative
MHC	—	Major Histocompatibility Complex
MIMBD	—	Meet. on the Interconnection of Mol. Biol. Databases
MIT	—	Massachusetts Institute of Technology
MMRF	—	Marshfield Medical Research Foundation
MNBWS	—	Miami Nature Biotechnology Winter Symposium

MOD	—	March of Dimes
MOU	—	Memorandum of Understanding
MRC	—	Medical Research Council
MS	—	Mass Spectrometry
NACHGR	—	Natl. Advisory Council for Human Genome Res.
NAPBC	—	Natl. Action Plan on Breast Cancer
NAS	—	Natl. Academy of Sciences
NASA	—	Natl. Aeronautics and Space Administration
NBAC	—	Natl. Bioethics Advisory Commission
NCGR	—	National Center for Genome Resources
NCI	—	Natl. Cancer Institute
NCSA	—	National Center for Supercomputing Applications
NCSL	—	National Conference of State Legislatures
NCSU	—	North Carolina State University
NEH	—	National Endowment for the Humanities
NFCR	—	National Foundation for Cancer Research
NFID	—	National Foundation for Infectious Diseases
NHGRI	—	Natl. Human Genome Res. Inst.
NICHD	—	National Institute of Child Health and Human Development
NIGMS	—	National Institute of General Medical Sciences

NIH	—	National Institutes of Health
NIGMS	—	Natl. Institute of General Medical Sciences
NIST	—	Natl. Inst. of Standards and Technology
NLGLP	—	National Laboratory Gene Library Project
NLM	—	National Library of Medicine
NMHCC	—	National Managed Health Care Congress
NORD	—	National Organization for Rare Disorders
NRC	—	National Research Council
NRSA	—	National Research Service Award
NSF	—	Natl. Sci. Foundation
NSGC	—	National Society of Genetic Counselors
NSTA	—	National Science Teachers Association
NYAS	—	New York Academy of Science
OAINN	—	Ohio Aerospace Institute Neural Networks
OBER	—	Office of Biological and Environmental Research
OHER	—	Office of Health and Environmental Research
OMIM	—	Online *Mendelian Inheritance in Man*
OPRR	—	Office of Protection from Res. Risks
ORAU	—	Oak Ridge Associated Universities
ORISE	—	Oak Ridge Inst. for Science and Education

ORF	—	open reading frame
ORNL	—	Oak Ridge National Laboratory
OSU	—	Oregon State University
OTA	—	Office of Technology Assessment
PACHG	—	Program Advisory Committee on the Human Genome
PCR	—	Polymerase Chain Reaction
PFGE	—	Pulsed-field gel electrophresis
PG	—	Plant Genome
PIR	—	Protein Info. Resource
PNNL	—	Pacific Northwest National Laboratory
PRIM&R	—	Public Responsibility in Medicine & Research
PSC	—	Pittsburgh Supercomputing Center
QDFM	—	Quantitative DNA Fiber Mapping
RAPD	—	Random Amplified Polymorphic DNA
RARA	—	Retinoic Acid Receptor
RECOMB	—	Conference on Computational Molecular Biology
RFLP	—	Restriction Fragment Length Polymorphism
RH	—	Radiation Hybrid
RLGS	—	Restriction Landmark Genomic Scanning
SASE	—	Sample Sequencing
SBH	—	Sequencing by Hybridization
SBIR	—	Small Business Innovation Research
SCAN	—	Sequence Comparison ANalysis Program

SCE	—	School of Continuing Education
SCI	—	Society of Chemical Industry
SCW	—	Single-chromosome Workshop
SDC	—	San Diego Conference
SERGG	—	Southeast Regional Genetics Group
SHOM	—	Sequencing by Hybridization on Matrices
SIAM	—	Society for Industrial and Applied Mathematics
SIB	—	Society for Industrial Biology
SIM	—	Society for Industrial Microbiology
SIMS	—	Societal Institute of the Mathematical Sciences
SIVB	—	Society for In Vitro Biology
SNL	—	Sandia National Laboratory
SNP	—	Single Nucleotide Polymorphism
snRNP	—	Small Nuclear Ribonucleo Protein Particle
SQL	—	Standard Query Language
SSLP	—	Single-sequence Length Polymorphism
SSR	—	Simple Sequence Repeat
STM	—	Scanning Tunneling Microscope
STRP	—	Short Tandem Repeat Polymorphism
STS	—	Sequence-tagged Site
SULS	—	Suffolk Univ. Law School
TCR	—	T-cell receptor
TIGR/NIST	—	The Inst. for Genomic Res./Natl. Inst. of Standards and Technol.

UC	—	University of California
UCB	—	University of California, Berkeley
UCLA	—	University of California, Los Angeles
UICM	—	Univ. of Illinois College of Medicine
UMBC	—	University of Maryland, Baltimore County
UNC	—	University of North Carolina at Chapel Hill
UMDS	—	United Medical & Dental Schools (Univ. of London)
UNESCO	—	United Nations Educational, Scientific, and Cultural Organization
USDA	—	U.S. Dept. of Agriculture
UT	—	University of Tennessee
UW	—	University of Washington
WCCS	—	Workshop on Complete cDNA Sequencing
WHS	—	Wolf-Hirschhorn syndrome
WLMG	—	Wellcome Laboratory for Molecular Genetics
WSES	—	World Scientific and Engineering Society
WWW	—	World Wide Web
XIST	—	candidate gene for X-inactivation center
YAC	—	yeast artificial chromosome

List of Glossary of Terms

Actin filament (microfilament): Helical protein filament formed by the polymerization of globular actin molecules. A major constituent of the cytoskeleton of all eucaryotic cells and part of the con- tractile apparatus of skeletal muscle.

Adaptive radiation: The evolution of new species or sub-species to fill unoccupied ecological niches.

Aerobe: A microorganism that grows in the presence of oxygen. See Anaerobe.

Agarose gel electrophoresis: A matrix composed of a highly purified form of agar that is used to separate larger DNA and RNA molecules ranging 20,000 nucleotides.

Algorithm: A set of well-defined mathematical rules or operations for solving a problem in a finite number of steps.

Alleles: Alternate forms of a gene or DNA sequence, which occur on either of two homologous chromosomes in a diploid organism.

Alternative mRNA splicing: The inclusion or exclusion of different exons to form different mRNA transcripts.

Ampicillin (beta-lactamase): An antibiotic derived from penicillin that prevents bacterial growth by interfering with cell wall synthesis.

Amplify: To increase the number of copies of a DNA sequence, in vivo by inserting into a cloning vector that replicates within a host cell, or in vitro by polymerase chain reaction (PCR).

Anaerobe: An organism that grows in the absence of oxygen. See Aerobe.

Analyte: A chemical species targeted for qualitative or quantitative analysis.

Anneal: The pairing of complementary DNA or RNA sequences, via hydrogen bonding, to form a double-stranded polynucleotide. Most often used to describe the binding of a short primer or probe.

Antibiotic resistance: The ability of a microorganism to produce a protein that disables an antibiotic or prevents transport of the antibiotic into the cell.

Antibiotic: A class of natural and synthetic compounds that inhibit the growth of or kill other microorganisms.

Antibody: An immunoglobulin protein produced by B-lymphocytes of the immune system that binds to a specific antigen molecule.

Anticodon: A nucleotide base triplet in a transfer RNA molecule that pairs with a complementary base triplet, or codon, in a messenger RNA molecule. See Codon, Messenger RNA, RNA.

Antigen: Any foreign substance, such as a virus, bacterium, or protein, that elicits an immune response by stimulating the production of antibodies.

Antigenic determinant: A surface feature of a microorganism or macromolecule, such as a glycoprotein, that elicits an immune response.

Antigenic switching: The altering of a microorganism's surface antigens through genetic rearrangement, to elude detection by the host's immune system.

Antimicrobial agent: Any chemical or biological agent that harms the growth of microorganisms.

Anti-oncogene: See Recessive oncogene.

Antisense DNA: DNA strand that is complementary (opposite) to the functional gene. Antisense DNA can block the function of the normal sense DNA. This method was used in the construction of the Flavr Savr tomato.

Antisense RNA: A complementary RNA sequence that binds to a naturally occurring (sense) mRNA molecule, thus blocking its translation.

Aromatic: A term used to describe cyclic pi-bonded structures of special stability.

Asexual reproduction: Nonsexual means of reproduction which can include grafting and budding.

Autoproductivity: The ability of a system, under external control, to automatically produce an identical copy of itself.

Bacillus: A rod-shaped bacterium.

Backcross: Crossing an organism with one of its parent organisms.

Bacteriocide: A class of antibiotics that kills bacterial cells.

Bacteriophage (phage or phage particle): A virus that in- fects bacteria. Altered forms are used as vectors for cloning DNA.

Bacteriophage (phage): Any virus that infects bacteria. They were the first organisms used for the study of molecular genetics and are now widely used as cloning vectors. (From Greek *phagein*, to eat.)

Bacteriorhodopsin: Pigmented protein found in the plasma membrane of a salt-loving bacterium, *Halobacterium halobium*; it pumps protons out of the cell in response to light.

Bacteriostat: A class of antibiotics that prevents growth of bacterial cells.

Bacterium (plural bacteria): Common name for any member of the diverse group of procaryotic organisms. Most are single cells, but multicellular forms also exist.

Bacterium: A single-celled, microscopic prokaryotic organism: a single cell organism without a distinct nucleus.

Band: In gels and blots, bands are visible indications of a particular fragment of a certain size. There may be one to many bands per lane.

Banoprism: Scientists at Northwestern University have created a nanoparticle with a new shape that could be a useful tool in the race to detect biological threats. The nanoprism, which resembles a tiny Dorito, exhibits unusual optical properties that could be used to improve biodetectors, allowing them to test for a far greater number of biological warfare agents or diseases at one time.

Banoscience: The study of phenomena and manipulation of materials at atomic, molecular and macromolecular scales, where properties differ significantly from those at a larger scale. Nanoscience is primarily the extension of existing sciences into the realms of the extremely small (*nanomaterials, nanochemistry, nanobio, nanophysics*, etc.) while *nanoengineering* represents the extension of the engineering fields into the nano- scale realm (*nanofabrication, nanodevices*, etc.). The exponential growth of nanoscience is largely due to the development of new instruments and related techniques that are used to "routinely" probe and manipulate material at the atomic and molecular level. *Scanning probe microscopies,* analytical electron- beam techniques, epitaxial growth facilities, and *synchrotron* radiation sources are all opening huge opportunities.

Banoshells: Procedures that target cancer cells while leaving normal cells untouched, patients controlling the release of medicine in their bodies with an infrared light, and medical test results produced in seconds rather than days – these are three new medical technologies currently being tested by nanotechnology researchers at Rice University. ... The research focuses on nanoshells, a new type of nanoparticle invented by Naomi Halas, Rice professor of electrical and computer engineering and chemistry. Nanoshells are layered nanoparticles whose ability to manipulate light and color can be designed into the nanoparticle by varying the thickness of the nanoparticle's layers.

Basal body: Short cylindrical array of microtubules plus their associated proteins found at the base of a eucaryotic cell cilium or flagellum. Serves as a nucleation site for the growth of the axoneme. Closely similar in structure to a centriole.

Basal lamina (plural basal laminae): Thin mat of extracellular matrix that separates epithelial sheets, and many types of cells such as muscle cells or fat cells, from connective tissue. Sometimes called a basement membrane.

Base pair (bp): A pair of complementary nitrogenous bases in a DNA molecule—adenine-thymine and guanine-cytosine. Also, the unit of measurement for DNA sequences. In DNA, there are four possible bases:

cytosine (C), guanine (G), adenine (A), and thymine (T). Cytosine and thymine are pyrimidine bases; adenine and guanine are purine bases. Cytosine is complementary to guanine while adenine is complementary to thymine. If one strand of DNA has the sequence ATTGC then the complementary strand will be TAACG. Two complementary bases constitute a base pair. In RNA, thymine is replaced by uracil.

Beta-DNA: The normal form of DNA found in biological systems, which exists as a right-handed helix.

Beta-Lactamase: Ampicillin resistance gene.

Binding energy: The reduction in the free energy of a system that occurs when a ligand binds to a receptor. Generally used to describe the total energy required to remove something, or to take a system apart into its constituent particles—for example, to separate two atoms from one another, or to separate an atom into electrons and nuclei.

Bio-assemblies or Biomolecular Assemblies: Containing several protein units, DNA loops, lipids, various ligands, etc.

Bioaugmentation: Increasing the activity of bacteria that decompose pollutants; a technique used in bioremediation.

Biochauvinism: The prejudice that biological systems have an intrinsic superiority that will always give them a monopoly on self-reproduction and intelligence.

Biocompatible coated materials: Biocompatible materials usually used in dental and bone implants that enhance biologic fixation, thereby increasing the bond strength between the coated material and bone, and minimize possible biological effects that may result from the implant itself. MeSH, 1999

Biocompatible materials: Synthetic or natural materials, other than drugs, that are used to replace or repair any body tissue or bodily function. MeSH, 1973

Biodefense: The need for improved national defenses against biological attacks are urgently needed. A real-time reporting system needs to be developed, whereby officials can be informed about an emerging threat before an agent has a chance to affect thousands of people. The development of integrated systems for detecting and monitoring biological agents is instrumental to this goal. The development of vaccines and novel detection platforms will enable preparedness, while also benefiting human health.

Biodiversity: The wide diversity and interrelatedness of earth organisms based on genetic and environmental factors.

Biodynotics Biologically Inspired Multifunctional Dynamic Robotics: A multidisciplinary, multi-pronged approach with far reaching impact on robotic capabilities for national security applications. Biologically Inspired Multifunctional Dynamic Robotics: BIODYNOTICS will explore the following areas: DYNAMIC MOBILITY, BEHAVIOR, INTEGRATION Biological Sciences, Defense Sciences Office, DARPA.

Bioengineering: Is rooted in physics, mathematics, chemistry, biology, and the life sciences. It is the application of a systematic, quantitative, and integrative way of thinking about and approaching the solutions of problems important to biology, medical research, clinical proactive, and population studies. The NIH Bioengineering Consortium agreed on the following definition for bioengineering research on biology, medicine, behavior, or health recognizing that no definition could completely eliminate overlap with other research disciplines or preclude variations in interpretation by different individuals and organizations. Integrates physical, chemical, or mathematical sciences and engineering principles for the study of biology, medicine, behavior, or health. It advances fundamental concepts, creates knowledge for the molecular to the organ systems levels, and develops innovative biologics, materials, processes, implants, devices, and informatics approaches for the prevention, diagnosis, and treatment of disease, for patient rehabilitation, and for improving health.

Bioenrichment: Adding nutrients or oxygen to increase microbial breakdown of pollutants.

Biofabrication: Includes nanoparticle delivery systems, biomaterials, tissue engineering, implants and prostheses. Using biological processes to synthesize and manufacture chemicals and materials of high value to the Department of Defense. Biological processes are characterized by: low energy barriers (~10Kcal, high temperatures/pressures not required); high reaction-, regio- and stereo- specificity (protection/ deprotection wastes and catalyst poisoning avoided); spatio-temporal control of materials synthesis of defined composition and size with angstrom- level precision; and local control of the dielectric environment (largely eliminating the need for toxic solvents). Potential target materials and chemicals include composites with enhanced mechanical properties, ultra-low k dielectrics and thermoelectrics, optoelectronic materials, photonic devices (waveguides and logic elements), electronic materials (e.g., GaN, InGaN, AlGaN), elastomers, energetic materials, and adhesives. Biofabrication, DARPA.

Biologics: Agents, such as vaccines, that give immunity to diseases or harmful biotic stresses.

Biomechanics: Mechanical structures of living organisms.

Biomedical Nanotechnology: (1) the comprehensive monitoring, control, construction, repair, defense, and improvement of all human biological systems, working from the molecular level, using engineered nanodevices and nanostructures; (2) the science and technology of diagnosing, treating, and preventing disease and traumatic injury, of relieving pain, and of preserving and improving human health, using molecular tools and molecular knowledge of the human body; (3) the employment of molecular machine systems to address medical problems, using molecular knowledge to maintain and improve human health at the molecular scale. (4) A rapidly expanding field that includes many potential technologies and approaches. The key to this definition is that phenomena and materials at the nanometer scale are known to have properties that are uniquely attributable to that scale length. Nanomedicine could similarly be defined [as nanotechnology] as the design, synthesis, or application of materials, devices, or technologies in the nanometer-scale for the basic understanding, diagnosis, and / or treatment of disease. Canadian Institute of Health Research, Regenerative Medicine and Nanomedicine, RFA, 2003. The monitoring, repair, construction and control of human biological systems at the molecular level, using engineered nanodevices and nanostructures. (5) the comprehensive monitoring, control, construction, repair, defense, and improvement of all human biological systems, working from the molecular level, using engineered nanodevices and nanostructures; (6) the science and technology of diagnosing, treating, and preventing disease and traumatic injury, of relieving pain, and of preserving and improving human health, using molecular tools and molecular knowledge of the human body; (7) the employment of molecular machine systems to address medical problems, using molecular knowledge to maintain and improve human health at the molecular scale. The use of molecular-scale devices (nanotechnology) to repair damage and boost the immune system. Vapour grown carbon fibres are obtained as shown in this SEM image. The diameters of these fibres can vary from 100 nm to 500 nm. As part of the National Institutes of Health (NIH) Roadmap for Medical Research [nihroadmap.nih. gov], the NIH will establish a handful of nanomedicine centers. These centers will be staffed by a highly interdisciplinary scientific crew including biologists, physicians, mathematicians, engineers and computer scientists. Research conducted over the first few years will be spent gathering extensive information about how molecular machines are built. A key activity during this time will be the development of a new kind of vocabulary, or lexicon, to define biological parts and processes in engineering terms. Once researchers have completely catalogued the interactions between and within molecules, they can begin to look for patterns and a higher order of connectedness than is possible to identify with current experimental methods. Mapping these networks and understanding how they change over time will be a crucial step toward helping scientists understand nature's rules of biological design. Understanding these rules will, in many years' time, enable researchers to use this information to address biological issues in unhealthy cells.

Biomedical polymers: More and more therapeutic problems are relevant to the use of polymer- based therapeutic aids for a limited period of time, namely the healing time related to the outstanding capacity of living systems to self- repair ... After healing the remaining prosthetic materials or devices become foreign residues or wastes that have to be eliminated from the body. Nowadays, biocompatible polymers that can degrade in the body are developed. The degradation and the elimination of degradation by-products depend on rather complex phenomena that are presently reflected inconsistently by terms issued from the tradition because each domain has developed its own terminology almost independently. This is a source of misunderstandings, confusions and misperceptions among scientists, surgeons, pharmacists, journalists and politicians, the situation being increased by the introduction of degradable polymers in plastic waste management and environmental protection. Therefore, it is urgent to reflect the various phenomena by specific terms, harmonize and enforce their use by the people active in the biomedical, pharmacological and environmental fields, and, last but not least by the publishing media and journalists. IUPAC, Terminology for biomedical (therapeutic) polymers.

BioMEMS Biological MicroElectro Mechanical Systems: Highlights the technical advances in the field that are leading a revolution in medicine, and creating a new generation of analytical devices for medical diagnostics. The meeting will encompass technology developments in micro & nano drug delivery, interface of nanotech and tissue engineering, microfluidics, and miniaturized total analysis systems (microTAS), biosensors, innovations in mass spec, and nanoscale imaging *BioMEMS and Nanotech World*, Aug. 16- 17, 2004, Washington DC

BioMEMS: MEMS used in medicine, that use microchips.

Biomimetic Chemistry: Knowledge of biochemistry, analytical chemistry, polymer science, and biomimetic chemistry is linked and applied to research in designing new molecules, molecular assemblies, and macromolecules having biomimetic functions. These new bio-related materials of high performance, including, for example, enzyme models, synthetic cell membranes, and biodegradable polymers, are prepared, tested, and constantly improved in this division for industrial scale production.

Biomimetic materials: Materials fabricated by BIOMIMETICS techniques, i.e., based on natural processes found in biological systems. [MeSH 2003] (2) Materials that imitate, copy, or learn from nature.

Biomimetic: Imitating, copying, or learning from nature. Nanotechnology already exists in nature; thus, nanoscientists have a wide variety of components and tricks already available. An interdisciplinary field in materials science, ENGINEERING, and BIOLOGY, studying the use of biological principles for synthesis or fabrication of BIOMIMETIC MATERIALS. There is a need to develop the next generation of restorative materials and medical implants. New avenues of scientific inquiry may enable the development of biomaterials that are safe, reliable, "smart", long-lasting, and perform ideally in their respective biological environments.... Over the last few years biomimetics and tissue engineering have emerged as a new vision in the field of tissue and organ repair and restoration. Biomimetics and tissue engineering are interdisciplinary fields that combine information from the study of biological structures and their functions with physics, mathematics, chemistry and engineering for the generation of new materials, tissues and organs. In the area of craniofacial, oral and dental principles from biomimetics and tissue engineering are applied to developing dental and facial implants, new polymers for guided tissue regeneration used in treating periodontal disease and bone and connective tissue defects, coral- based hydroxyapatite replicas for reconstruction of alveolar ridges and other osseous defects, temporomandibular joint (TMJ) and other joint prostheses, formation of bone matrix substitutes, and artificial replicas of bone, skin, and mucosa. The term biomimetics was coined in 1972 in the context of artificial enzymes; it might be defined broadly as "the abstraction of good design from nature". It is a fact of everyday life that nature has managed to built materials and 'devices' with breathtaking functionality, heterogeneity and stability by using a comparatively limited number of building blocks (the whole range of synthetic materials is restricted to man- made engineering). The basic concepts of nature are often simple; it is the way in which building blocks and materials are arranged that results in functionality. Among the most simple and abundant themes of nature is self- assembly: lipids assemble in sheets to form cell membranes, proteins assemble into functional enzymes, cellular 'sensors', fibers, or virus coats, and DNA assembles in double strands to provide the very basis for live: replication. Biomimetics Study of the structure and function of biological substances to make artificial products that mimic the natural ones.

Biomotors: Driven by energy sources such as *adenosine triphosphate (ATP)* for chemical transduction and other processes. These biomotors are considered to be biomolecular and are discussed in the body of this report, but strictly speaking they do not conform to the panel's definition of self- assembly.

Bionanotechnology: Includes molecular motors, biomaterials, single molecule manipulation technologies, biochip technologies, etc.

Biopolymeroptoelectromechanical Systems [BioPOEMS]: Combining optics and microelectromechanical systems, and used in biological applications.

Biopolymers: Macromolecules (including proteins, nucleic acids and polysaccharides) formed by living organisms.

Bioremediation: The use of microorganisms to remedy environmental problems. See Bioaugmentation, Bioenrichment.

Biorobotics: Our research focuses on the role of sensing and mechanical design in motor control, in both robots and humans. This work draws upon diverse disciplines, including biomechanics, systems analysis, and neurophysiology. The main approach is experimental, although analysis and simulation play important parts. In conjunction with industrial partners, we are developing applications of this research in biomedical instrumentation, teleoperated robots, and intelligent sensors.

Biosensor: The term "biosensor" is a general designation that denotes either a sensor to detect a biological substance or a sensor which incorporates the use of biological molecules such as antibodies or enzymes. Biosensors are a subcategory of chemical sensors.

Biostasis: A condition in which an organism's cell and tissue structure are preserved, allowing later restoration by cell repair machines. Applicable to cryonics.

Biotechnology: The scientific manipulation of living organisms, especially at the molecular genetic level, to produce useful products. Gene splicing and use of recombinant DNA (rDNA) are major techniques used. Technology for working with biological systems. Includes genetic engineering, human and veterinary medicine, crop and animal breeding, diagnostics, pharmaceuticals, forensics, etc. Narrow sense: Genetic engineering.

Biotic stress: Living organisms which can harm plants , such as viruses, fungi, and bacteria, and harmful insects. See Abiotic stress.

Biovorous: From "biovore;" an organism capable of converting biological material into energy for sustenance.

Bipolar-junction transistor: Transistor with n-type and p-type semiconductors having base-emitter and collector-base junctions.

Blocks Substitution Matrix: A substitution matrix in which scores for each position are derived from *observations* of the frequencies of substitutions in blocks of local alignments in related proteins. Each matrix is tailored to a particular evolutionary distance. In the BLOSUM62 matrix, for example, the alignment from which scores were derived was created using sequences sharing no more than 62% identity. Sequences more identical than 62% are represented by a single sequence in the alignment so as to avoid over-weighting closely related family members. (Henikoff and Henikoff)

Blot, northern: A pattern of RNA fragments transferred to a nitrocellulose membrane from a gel. The gel has undergone electrophoresis to separate fragments according to size. The RNA fragments are arranged in different lanes for each sample. Each lane contains bands which are fragments of different sizes. Radioactive probes are often used to visualized particular bands by autoradiography.

Blot, southern: A pattern of DNA fragments transferred to a nitrocellulose membrane from a gel. The gel has undergone electrophoresis to separate fragments according to size. The DNA fragments are arranged in different lanes for each sample. Each lane contains bands which are fragments of different sizes. Radioactive probes are often used to visualized particular bands by autoradiography.

Blot, western: A pattern of proteins transferred to a nitrocellulose membrane from a gel. The gel has undergone electrophoresis to separate fragments according to size. The proteins are arranged in different lanes for each sample. Each lane contains bands which are fragments of different sizes. Labeled antibody probes are often used to visualize particular bands.

Bone substitutes: Synthetic or natural materials for the replacement of bones or bone tissue. They include hard tissue replacement polymers, natural coral, hydroxyapatite, beta- tricalcium phosphate, and various other biomaterials. The bone substitutes as inert materials can be incorporated into surrounding tissue or gradually replaced by original tissue.

Carcinogen: A substance that induces cancer.

Carcinoma: A malignant tumor derived from epithelial

tissue, which forms the skin and outer cell layers of internal organs.

Catalyst: A substance that promotes a chemical reaction by lowering the activation energy of a chemical reaction, but which itself remains unaltered at the end of the reaction.

Catalytic antibody (abzyme): An antibody selected for its ability to catalyze a chemical reaction by binding to and stabilizing the transition state intermediate.

Catalytic RNA (ribozyme): A natural or synthetic RNA molecule that cuts an RNA substrate.

Cation: A positively charged ion.

cDNA library: A library composed of complementary copies of cellular mRNAs.

cDNA: DNA synthesized from an RNA template using reverse transcriptase. Complementary DNA to a particular RNA fragment.

Cell pharmacology: Delivery of drugs by medical nanomachines to exact locations in the body.

Cell Repair Machine: A system including nanocomputers and molecular scale sensors and tools, programmed to repair damage to cells and tissues. Molecular and nanoscale machines with sensors, nanocomputers and tools, programmed to detect and repair damage to cells and tissues, which could even report back to and receive instructions from a human doctor if needed.

Cell surgery: In medical nanorobotics, modifying cellular structures using medical nanomachines. Modifying cellular structures using medical nanomachines.

Cell: A membrane-bound unit, typically microns in diameter. All plants and animals are made up of one or more cells (trillions, in the case of human beings). In general, each cell of a multicellular organism contains a nucleus holding all of the genetic information of the organism. (2) A small structural unit, surrounded by a membrane, making up living things.

Cellular oncogene (proto-oncogene): A normal gene that when mutated or improperly expressed contributes to the development of cancer.

Centers of origin: Usually the location in the world where the oldest cultivation of a particular crop has been identified.

Central dogma: Francis Crick's seminal concept that in nature genetic information generally flows from DNA to RNA to protein.

Centrifugation: Separating molecules by size or density using centrifugal forces generated by a spinning rotor. G forces of several hundred thousand times gravity are generated in ultracentrifugation.

Centromere: The central portion of the chromosome to which the spindle fibers attach during mitotic and meiotic division.

Chemotherapy: A treatment for cancers that involves administering chemicals toxic to malignant cells.

Chloramphenicol: An antibiotic that interferes with protein synthesis.

Chromatid: Each of the two daughter strands of a duplicated chromosome joined at the centromere during mitosis and meiosis.

Chromosome walking: Working from a flanking DNA marker, overlapping clones are successively identified that span a chromosomal region of interest.

Chromosome: A single DNA molecule, a tightly coiled strant of DNA, condensed into a compact structure in vivo by complexing with accessory histones or histone-like proteins. Chromosomes exist in pairs in higher eukaryotes. A linear structure in the nucleus of plants and animals that is visible in light microscopy when stained. The chromosome is a single, long, linear molecule of DNA and associated proteins. Bacteria have a single circular chromosome; other organisms may have many linear chromosomes.

Cistron: A DNA sequence that codes for a specific polypeptide; a gene. See DNA, Gene.

Clone: An exact genetic replica of a specific gene or an entire organism. See Cloning.

Cloning vector: A DNA molecule capable of autonomous replication within the cloning host cell (e.g *E. coli*). The vectors contain restriction enzyme sites for insertion of foreign DNA. Cloning vectors are derived from bacterial plasmids, bacteriophages, or viruses.

Cloning: The mitotic division of a progenitor cell to give rise to a population of identical daughter cells or clones.

Coat protein (capsid): The coating of a protein that enclosed the nucleic acid core of a virus.

Codon: A group of three nucleotides that specifies addition of one of the 20 amino acids during translation of an mRNA into a polypeptide. Strings of codons form genes and strings of genes form chromosomes. Set of three nucleotides that specify a particular amino acid during protein synthesis.

Coenzyme (cofactor): An organic molecule, such as a vitamin, that binds to an enzyme and is required for its catalytic activity.

Colony: A group of identical cells (clones) derived from a single progenitor cell.

Combinatorial materials design: Uses computing power (sometimes together with massive parallel experimentation) to screen many different materials possibilities to optimize properties for specific applications (e.g., catalysts, drugs, optical materials).

Commensalism: The close association of two or more dissimilar organisms where the association is advantageous to one and doesn't affect the other(s). See Parasitism, Symbiosis.

Competency: An ephemeral state, induced by treatment with cold cations, during which bacterial cells are capable of uptaking foreign DNA.

Complementary DNA or RNA: The matching strand of a DNA or RNA molecule to which its bases pair.

Complementary nucleotides: Members of the pairs adenine-thymine, adenine-uracil, and guaninecytosine that have the ability to hydrogen bond to one another.

Concatemer: A DNA segment composed of repeated sequences linked end to end.

Configuration space: A mathematical space describing the three-dimensional configuration of a system of particles (e.g., atoms in a nanomechanical structure) as a single point; the configuration space for an N particle system has $3N$ dimensions.

Conformation: The shape of a molecule. A molecular geometry that differs from other geometries chiefly by rotation about single or triple bonds; distinct conformations (termed con formers) are associated with distinct potential wells. Typical biomolecules and products of organic synthesis can interconvert among many conformations; typical diamondoid structures are locked into a single potential well, and thus lack conformational flexibility.

Conjugation: The joining of two bacteria cells when genetic material is transferred from one bacterium to another.

Constitutive promoter: An unregulated promoter that allows for continual transcription of its associated gene.

Contiguous (contig) map: The alignment of sequence data from large, adjacent regions of the genome to produce a continuous nucleotide sequence across a chromosomal region.

Coulomb [C]: Measure of electrical charge: 1 C is an amount of charge equal to that of about 6.24x1018 electrons.

Coupling: A connection between more than one object or energy pathway so that together they function as a single unit.

Cross-hybridization: The hydrogen bonding of a single-stranded DNA sequence that is partially but not entirely complementary to a singlestranded substrate. Often, this involves hybridizing a DNA probe for a specific DNA sequence to the homologous sequences of different species.

Crossing-over: The exchange of DNA sequences between chromatids of homologous chromosomes during meiosis.

Cross-pollination: Fertilization of a plant from a plant with a different genetic makeup.

Crosstalk: Electromagnetic noise transmitted between leads or circuits in close proximity to each other.

Crybiology: The science of biology at low temperatures; research in cryobiology has made possible the freezing and storing of sperm and blood for later use.

Crystal Lattice: The regular three-dimensional pattern of atoms in a crystal.

Crystal structure: For crystalline materials, the manner in which atoms or ions are arrayed in space. It is defined in terms of the unit cell geometry and the atom positions within the unit cell.

Crystallescence: In medical nanorobotics, the crystallization of solid solute that is offloaded by nanorobot sorting rotors at a concentration that exceeds the solvation capacity of the surrounding solvent.

Culture: A particular kind of organism growing in a laboratory medium.

Curie temperature (also Curie point) (Tc): The temperature above which a ferromagnetic or ferrimagnetic material becomes paramagnetic. For iron the Curie point is 760oC and for nickel 356oC.

Current [A]: Measure of rate of flow of electric charge: a one-ampere current is a flow of 1 C of charge per second.

Cutoff: Condition in a diode or bipolar junction transistor in which the potential across a p-n junction prevents current flow.

Cyclic AMP (cyclic adenosine monophosphate): A second messenger that regulates many intracellular reactions by transducing signals from extracellular growth factors to cellular metabolic pathways.

Cyclic: A structure is termed cyclic if its covalent bonds form one or more rings.

Cytogenetics: Study that relates the appearance and behavior of chromosomes to genetic phenomenon.

Dalton: A unit of measurement equal to the mass of a hydrogen atom, 1.67 x 10E-24 gram/L (Avogadro's number).

Denature: To induce structural alterations that disrupt the biological activity of a molecule. Often refers to breaking hydrogen bonds between base pairs in double-stranded nucleic acid molecules to produce in single-stranded polynucleotides or altering the secondary and tertiary structure of a protein, destroying its activity.

Density gradient centrifugation: High-speed centrifugation in which molecules "float" at a point where their density equals that in a gradient of cesium chloride or sucrose.

Diabetes: A disease associated with the absence or reduced levels of insulin, a hormone essential for the transport of glucose to cells.

Dideoxynucleotide (didN): A deoxynucleotide that lacks a 3' hydroxyl group, and is thus unable to form a 3'-5' phosphodiester bond necessary for chain elongation. Dideoxynucleotides are used in DNA sequencing and the treatment of viral diseases.

Digest: To cut DNA molecules with one or more restriction endonucleases.

Diploid cell: A cell which contains two copies of each chromosome. See Haploid cell.

Directional cloning: DNA insert and vector molecules are digested with two different restriction enzymes to create noncomplementary sticky ends at either end of each restriction fragment. This allows the insert to be ligated to the vector in a specific orientation and prevents the vector from recircularizing.

DNA: DNA is the recipe for life. DNA is a molecule found in the nucleus of every cell and is made up of 4 nucleotides Adenine (A), Guanine (G), Thymine (T), and Cystosine (C). The order of the nucleotides in the DNA strand holds a code of information for the cell.

DNA (Deoxyribonucleic acid): An organic acid and polymer composed of four nitrogenous bases—adenine, thymine, cytosine, and guanine linked via intervening units of phosphate and the pentose sugar deoxyribose. DNA is the genetic material of most organisms and usually exists as a double-stranded molecule in which two antiparallel strands are held together by hydrogen bonds between adeninethymine and cytosine-guanine.

DNA (Deoxyriobonucleic Acid): DNA molecules are long chains consisting of four kinds of nucleotides; the order of these nucleotides encodes the information needed to construct protein molecules. These in turn make up much of the molecular machinery of the cell. DNA is the genetic material of cells.

DNA Chip: Also: Gene Chip and DNA Microchip. A purpose built microchip used to identify mutations or alterations in a gene's DNA.

DNA diagnosis: The use of DNA polymorphisms to detect the presence of a disease gene.

DNA diagnostics - miniaturization of: In the areas of sample preparation and assay, it is clear that miniaturization is key. To reduce the size of samples by a factor of 10 or greater, barriers in microfluidics, micromachining, robotics, microchemistry, nucleic acid chemistry, and surface chemistry must be overcome. To implement miniaturized protocols accurately and efficiently, substantial automation of the process will be required. In the development of miniaturized systems, it is essential that the system can be adapted for high levels of parallelization. Miniaturization poses significant technological risks. Currently, there exists .no universally accepted precedent for the handling, replication, amplification, or cloning of DNA in nanoliter volumes. Due to the size and charge of the DNA molecule, and the relative instability of many of the enzymes involved in the sample preparation processes, nanoliter and less volumes may pose substantial challenges. In addition, interactions of the biological molecules with the surfaces of the reaction chambers must be minimized. For some methodologies, it is not clear what the optimal sample will be, so substantial improvement in DNA fragmentation technologies or DNA cloning vectors may be required for the ultimate efficient application to diagnostics. Improvements in any of these areas are likely to be of value to other non-DNA based diagnostic applications such as antibody screening protocols and enzyme based diagnostics, because miniaturized robotic or micro- electro mechanical systems developed for DNA could be modified to be used for these purposes.

DNA fingerprint: The unique pattern of DNA fragments identified by Southern hybridization (using a probe that binds to a polymorphic region of DNA) or by polymerase chain reaction (using primers flanking the polymorphic region).

DNA polymorphism: One of two or more alternate forms (alleles) of a chromosomal locus that differ in nucleotide sequence or have variable numbers of repeated nucleotide units.

DNA Probes: A DNA or nucleic acid probe is a short strand of DNA that locates and binds to its complementary sequence in samples containing single strands of DNA or RNA enabling identification of specific sequences. Nucleic acid probe assays exploit the fundamental hybridization reaction that occurs spontaneously between two complementary DNA:DNA or DNA:RNA strands. As in immunoassays, detection of the hybrid requires that the probe be labeled. Various direct and indirect methods have been devised for the detection of the hybrid. Direct labeling involves attaching the label directly to the probe sequence; indirect labeling binds an antibody to the DNA:DNA or DNA:RNA hybrid. As in immunoassays, non-isotopically-labeled probes are preferred over radio-labeled probes primarily because of radiation hazards, disposal problems, and short reagent shelf life. In addition, the factors determining the detection limits of hybridization assays based on labeled probes are similar to those in immunoassays. Therefore, the development of a simple, inexpensive and sensitive direct detection system which eliminates the use of labels is highly desirable.

DNA Sequencing: There are two main classical methods for sequencing DNA: The first method, developed by Allan Maxam and Walter Gilbert, involves chemicals used to cleave the DNA at certain positions, generating a set of fragments that differ by one nucleotide. The second method, developed by Fred Sanger and Alan Coulson, involves enzymatic synthesis of DNA strands that terminate in a modified nucleotide. Analysis of fragments is similar for both methods and involves gel electrophoresis and autoradiography or fluorescence. The enzymatic method has largely replaced the chemical method as the technique of choice, although there are some situations where chemical sequencing can provide data more easily than the enzymatic method.

DNA, recombinant: DNA that has been cut and spliced

back together in a new sequence. The DNA may be from one organism or from more than one organism

DNA, repetitive: Fragments of DNA that appear in multiple copies in a single individual.

DNA: A molecule encoding genetic information, found in the cell's nucleus. Deoxyribose Nucleic Acid. DNA is a molecule used in biological systems to store genetic information. It has a long, double stranded structure with the two strands wrapped around each other to form a right handed helix. There are about 10 nucleotide pairs per helical turn. Each strand is composed of a sugar phosphate backbone and attached bases. It is connected to a complementary strand by hydrogen bonding (non- covalent) between paired bases, adenine (A) with thymine (T) and guanine (G) with cytosine (C). DNA is an abbreviation of DeoxyriboNucleic Acid. : It is the genetic 'blueprint' of all life on this planet. : DNA typing is often applied to Human Identification which has also been called DNA Fingerprinting. DNA Fingerprinting has been attributed to Alex Jeffries, who with his colleagues in 1985, discovered a sequence of nucleotides, subsequently termed short tandem repeats (STR) when comparing the DNA sequence of the human á-globin chromosomes isolated from different individuals.

DNA: Deoxyribonucleic acid, the molecule that stores genetic information. Composed of two complementary strands. See base pairs.

Dominant gene: A gene whose phenotype is when it is present in a single copy.

Dominant(-acting) oncogene: A gene that stimulates cell proliferation and contributes to oncogenesis when present in a single copy.

Dominant: An allele is said to be dominant if it expresses its phenotype even in the presence of a recessive allele. See Allele, Phenotype, Recessive.

Dormancy: A period in which a plant does not grow, awaiting necessary environmental conditions such as temperature, moisture, nutrient availability.

Double helix: Describes the coiling of the antiparallel strands of the DNA molecule, resembling a spiral staircase in which the paired bases form the steps and the sugar-phosphate backbones form the rails.

Double-stranded complementary DNA (dscDNA): A duplex DNA molecule copied from a cDNA template.

Downstream: The region extending in a 3' direction from a gene.

E: coli (Escherichia coli): A common intestinal bacterium that is widely used in genetic engineering as a host for a cloning vector. Some strains of E. coli are important foodborne pathogens. Lab strains are of the harmless variety.

Ecology: The study of the interactions of organisms with their environment and with each other.

Ecosystem: The organisms in a plant population and the biotic and abiotic factors which impact on them. See abiotic factors; Biotic factors.

Electrophoresis: The technique of separating charged molecules in a matrix to which is applied an electrical field.

Electroporation: A method for transforrning DNA, especially useful for plant cells, in which high voltage pulses of electricity are used to open pores in cell membranes, through which foreign DNA can pass.

Electroporation: Transformation technique that uses

electric fields to temporarily increase permeability of cells to foreign DNA.

Emergence: A complex whole created by simple parts, as in the brain where billions of neurons work individually, but collectively make up our consciousness and give us the ability to think, rationalize, and create.

Emulation: An absolutely precise simulation of something, so exact that it is equivalent to the original (for example, many computers emulate obsolete computers to run their programs). [AS] The Star Trek replicator is an example.

Enabling science and technologies: Areas of research relevant to a particular goal, such as nanotechnology. Also, technology that "enables" other technology to advance, such as the transistor enabled the computer chip revolution, as did photolithography.

Encapsidation: Process by which a virus' nucleic acid is enclosed in a capsid. See Coat protein.

Endocrine cell: Specialized animal cell that secretes a hormone into the blood; usually part of a gland, such as the thyroid or pituitary gland.

Endocytosis: Uptake of material into a cell by an invagination of the plasma membrane and its internalization in a membrane-bounded vesicle.

Endophyte: An organism that lives inside another.

Endoplasmic reticulum (ER): Labyrinthine, membrane-bounded compartment in the cytoplasm of eucaryotic cells, where lipids are synthesized and membrane-bound proteins are made.

Endosome: Membrane-bounded organelle in animal cells that carries materials newly ingested by endocytosis and passes many of them on to lysosomes for degradation.

Endothelium: Single sheet of highly flattened cells (endothelial cells) that forms the lining of all blood vessels. Regulates exchanges between the bloodstream and surrounding tissues and is usually surrounded by a basal lamina.

Energy [J]: Capacity for performing work or to cause heat flow. Like work itself, it is measured in Joules.

Environmental Protection Agency (EPA): The U.S. regulatory agency for biotechnology of microbes. The major laws under which the agency has regulatory powers are the Federal Insecticide, Fungicide, and Rodenticide Act (FIFRA); and the Toxic Substances Control Act (TSCA).

Enzymes: Proteins that control the various steps in all chemical reactions.

EPROM: Electrically Programmable Read-Only Memory — nonvolatile memory device.

Escherichia coli: A commensal bacterium inhabiting the human colon that is widely used in biology, both as a simple model of cell biochemical function and as a host for molecular cloning experiments.

Ethidium bromide: A fluorescent dye used to stain DNA and RNA. The dye fluoresces when exposed to UV light.

Eukaryote: An organism whose cells possess a nucleus and other membrane-bound vesicles, including all members of the protist, fungi, plant and animal kingdoms; and excluding viruses, bacteria, and blue-green algae. See Prokaryote.

Eurisko: A computer program developed by Professor

Douglas Lenat which is able to apply heuristic rules for performing various tasks, including the invention of new heuristic rules.

Event: An "event" in genetic engineering is the insertion of a particular piece of foreign DNA into the chromosome of the recipient. Insertion occurs in random locations, so each event is unique. The event can affect how a gene is expressed in the organism. Once an event occurs, the transgene can be passed to the next generation as a normally inherited gene.

Evolution: The long-term process through which a population of organisms accumulats genetic changes that enable its members to successfully adapt to environmental conditions and to better exploit food resources.

Exon: A DNA sequence that is ultimately translated into protein. See DNA.

Expectation value: The number of different alignents with scores equivalent to or better than S that are expected to occur in a database search by chance. The lower the E value, the more significant the score.

Express: To translate a gene's message into a molecular product.

Face-centered cubic: A crystal structure found in common elemental metals also FCC. Within the cubic unit cell, atoms are located at all corner and face-centered positions.

Fail-stop: Describes a component or subsystem that, in the event of a failure, produces no output (e.g., of material or data) rather than producing a damaged or incorrect output.

Fermi energy: The energy level in a solid at which the probability of finding an electron is 1/2. For a metal, the energy corresponding to the highest filled electron state in the valence band at 0 K.

Field-Effect Transistor (FET): Field-Effect Transistor — semiconductor device whose insulated gate electrode controls current flow.

FIFRA: The Federal Insecticide, Fungicide, and Rodenticide Act. See Environmental Protection Agency.

Fingerprint: A set of molecular markers sufficiently diverse to identify particular individuals with reasonable certainty.

Flanking region: The DNA sequences extending on either side of a specific locus or gene.

Food and Drug Administration (FDA): The U.S. agency responsible for regulation of biotechnology food products. The major laws under which the agency has regulatory powers include the Food, Drug, and Cosmetic Act; and the Public Health Service Act.

Fusion gene: A hybrid gene created by joining portions of two different genes (to produce a new protein) or by joining a gene to a different promoter (to alter or regulate gene transcription).

Gamete: A haploid sex cell, egg or sperm, that contains a single copy of each chromosome.

Ganglioside: Any glycolipid having one or more sialic acid residues in its structure. Found in the plasma membrane of eucaryotic cells and especially abundant in nerve cells.

Gel, agarose: A substance used to separate DNA or RNA fragments by size. Used for southern and northern blots.

Gel, polyacrylamide: A substance used to separate proteins by size. Used for western blots.

GEM: A genetically engineered microorganism.

Gene amplification: The presence of multiple genes. Amplification is one mechanism through which proto-oncogenes are activated in malignant cells.

Gene cloning: The process of synthesizing multiple copies of a particular DNA sequence using a bacteria cell or another organism as a host. See DNA, Host.

Gene expression: The process of producing a protein from its DNA- and mRNA-coding sequences.

Gene flow: The exchange of genes between different but (usually) related populations.

Gene frequency: The percentage of a given allele in a population of organisms. See Allele.

Gene gun: Transformation technique that uses accelerated particles coated with DNA to introduce foreign DNA into recipient plant.

Gene insertion: The addition of one or more copies of a normal gene into a defective chromosome.

Gene linkage: The hereditary association of genes located on the same chromosome.

Gene modification: The chemical repair of a gene's defective DNA sequence. See DNA.

Gene pool: The totality of all alleles of all genes of all individuals in a particular population.

Gene therapy: The introduction of new genes into individuals to cure diseases or genetic abnormalities.

Gene translocation: The movement of a gene fragment from one chromosomal location to another, which often alters or abolishes expression.

Gene: The basic unit of inheritance. A segment of DNA that codes for a particular protein. A locus on a chromosome that encodes a specific protein or several related proteins. It is considered the functional unit of heredity.

Genetic assimilation: Eventual extinction of a natural species as massive pollen flow occurs from another related species and the older crop becomes more like the new crop. See Gene flow.

Genetic code: The three-letter code that translates nucleic acid sequence into protein sequence. The relationships between the nucleotide base-pair triplets of a messenger RNA molecule and the 20 amino acids that are the building blocks of proteins. See Base pair, Nucleic acid, Nucleotide. The genetic information in DNA is encoded with four different nucleotide bases: A, C, G, and T. A set of three consecutive nucleotide bases constitutes a codon. A codon specifies a particular amino acid that is added during synthesis of a protein.

Genetic disease: A disease that has its origin in changes to the genetic material, DNA. Usually refers to diseases that are inherited in a Mendelian fashion, although noninherited forms of cancer also result from DNA mutation.

Genetic drift: Random variation in gene frequency from one generation to another.

Genetic engineering: Genetic engineering is the process of manually adding new DNA on a molecular level with the goal of adding one or more new traits that are not already found in that organism. For the purposes of this web site, genetic engineering includes recombinant DNA and gene splicing technologies.

Genetic linkage map: A linear map of the relative positions of genes along a chromosome. Distances are established by linkage analysis, which determines the frequency at which two gene loci become separated during chromosomal recombination. (

Genetic marker: A gene or group of genes used to "mark" or track the action of microbes.

Genetic modification: The selective and purposeful alteration of genes by humans. Genetic modification includes methods of incorporating new genes or traits into an organism through genetic engineering, mutagenesis and traditional breeding methods.

Genome: The genetic complement contained in the chromosomes of a given organism, usually the haploid chromosome state. The full chromosome set containing all the genes of a particular individual.

Genomic library: A library composed of fragments of genomic DNA.

Genomics: The study of the structure and function of genomes. Genomics usually involves high speed sequencing of the DNA and computer searches for sequences that code for genes.

Genotype: The sum total of the genetic information of an organism including the linkage relationships between genes. The genotype, modified by environment, determines the phenotype.

Genus: A category including closely related species. Interbreeding between organisms within the same category can occur.

Germ cell (germ line) gene therapy: The repair or replacement of a defective gene within the gamete-forming tissues, which produces a heritable change in an organism's genetic constitution.

GMO: Genetically modified organism. An organism that

has incorporated a functional foreign gene through recombinant DNA technology. The novel gene exists in all of its cells and is passed through to progeny. Same as transgenic.

Green revolution: Advances in genetics, petrochemicals, and machinery that culminated in a dramatic increase in crop productivity during the third quarter of the 20th century.

Growth curve: See Growth phase.

Growth factor: A serum protein that stimulates cell division when it binds to its cell-surface receptor.

Growth phase (curve): The characteristic periods in the growth of a bacterial culture, as indicated by the shape of a graph of viable cell number versus time.

GUI: Graphical User Interface — hardware, software, and firmware that produces the display on modern personal computers.

Guy Fawkes Scenario: If nanotechnology becomes widely available, it might become trivial for anyone to committ acts of terrorism (such as making nanomachines build a large amount of explosives under government buildings a la Guy Fawkes). This would either force strict control over nanotechnology (hard) or a decentralized mode of organization.

Hair cell: Specialized sensory epithelial cell in the ear with bundles of giant microvilli (stereocilia) protruding from its apical surface. Sound vibrations tilt the stereocilia, evoking an electrical change in the hair cell, which thus acts as a sound detector.

Hall effect: The phenomenon whereby a force is applied to a moving electron or hole by a magnetic field that is applied perpendicular to the direction of motion. The force direction is perpendicular to both the magnetic field and the particle motion directions.

Haploid cell: A cell containing only one set, or half the usual (diploid) number, of chromosomes.

Heme: An iron complex. Cyclic organic molecule containing an iron atom that carries oxygen in hemoglobin and carries an electron in cytochromes.

Hemophilia: An X-linked recessive genetic disease, caused by a mutation in the gene for clotting factor VIII (hemophilia A) or clotting factor IX (hemophilia B), which leads to abnormal blood clotting.

Herbicide resistance: Herbicide resistance is when plants become resistant to the negative effects of a particular herbicide formulation or are genetically engineered in order to have resistance to a particular herbicide. Herbicide-resistant crops can be sprayed with a particular herbicide, killing weeds growing in and around the crop plants, without being damaged.

Herbicide: Any substance that is toxic to plants; usually used to kill specific unwanted plants.

Heterochromatin: Dark-stained regions of chromosomes thought to be for the most part genetically inactive.

Heteroduplex: A double-stranded DNA molecule or DNA-RNA hybrid, where each strand is of a different origin.

Heterogeneous nuclear RNA (hnRNA): The name originally given to large RNA molecules found in the nucleus, which are now known to be unedited mRNA transcripts, or pre-mRNAs.

Homologous chromosomes: Chromosomes that have the same linear arrangement of genes—a pair of matching chromosomes in a diploid organism. See Chromosomes.

Homologous recombination: The exchange of DNA fragments between two DNA molecules or chromatids of paired chromosomes (during crossing over) at the site of identical nucleotide sequences.

Homozygote: An organism whose genotype is characterized by two identical alleles of a gene. See Allele, Genotype.

Host: An organism that contains another organism.

Human Genome Project: A project coordinated by the National Institutes of Health (NIH) and the Department of Energy (DOE) to determine the entire nucleotide sequence of the human chromosomes.

Human growth hormone (HGH, somatotrophin): A protein produced in the pituitary gland that stimulates the liver to produce somatomedins, which stimulate growth of bone and muscle.

Hybrid: The offspring of two parents differing in at least one genetic characteristic (trait). Also, a heteroduplex DNA or DNA-RNA molecule.

Hybridization, nucleic acid: Complementary strands of DNA or RNA will spontaneously match up with each other and bond together under the right conditions. This is called nucleic acid hybridization and is used when probing southern and northern blots.

Hybridization: The hydrogen bonding of complementary DNA and/or RNA sequences to form a duplex molecule.

Hybridoma: A hybrid cell, composed of a B Iymphocyte fused to a tumor cell, which grows indefinitely in tissue culture and is selected for the secretion of a specific antibody of interest.

Hydrogen bond: A relatively weak bond formed between a hydrogen atom (which is covalently bound to a nitrogen or oxygen atom) and a nitrogen or oxygen with an unshared electron pair.

Hydrolysis: A reaction in which a molecule of water is

added at the site of cleavage of a molecule to two products.

Imaging Contrast Agent: A molecule or molecular complex that increases the intensity of the signal detected by an imaging technique, including MRI and ultrasound. An MRI contrast agent, for example, might contain gadolinium attached to a targeting antibody. The antibody would bind to a specific target—a metastatic melanoma cell, for example – while the gadolinium would increase the magnetic signal detected by the MRI scanner.

Immortalization: Production of a cell line capable of an unlimited number of cell divisions. Can be the result of a chemical or viral transformation or of fusion with cells of a tumor line.

Immortalizing oncogene: A gene that upon transfection enables a primary cell to grow indefinitely in culture.

In situ: Refers to performing assays or manipulations with intact tissues.

In vivo: Refers to biological processes that take place within a living organism or cell.

Incomplete dominance: A condition where a heterozygous off- spring has a phenotype that is distinctly different from, and intermediate to, the parental phenotypes. See Heterozygote, Phenotype.

Initiation codon: The mRNA sequence AUG, coding for methionine, which initiates translation of mRNA.

Inositol lipid: A membrane-anchored phospholipid that transduces hormonal signals by stimulating the release of any of several chemical messengers.

insect resistance management: A set of strategies to reduce the frequency of resistance in crop pests. These strategies are widely referred to as Insect Resistance Management.

Insertion mutations: Changes in the base sequence of a DNA molecule resulting from the random integration of DNA from another source. See DNA, Mutation.

Insulator: A substance, object or material that does not conduct electricity. Material that conducts electricity very poorly.

Insulin: A peptide hormone secreted by the islets of Langerhans of the pancreas that regulates the level of sugar in the blood.

Integrated Circuit (IC): An electronic circuit consisting of many interconnected devices on one piece of semiconductor, typically into 10 millimeters on a side. ICs are the major building blocks of today's computers. Semiconductor circuit, typically on a very small silicon chip, containing microfabricated transistors, diodes, resistors, capacitors, etc.

Interferon: A family of small proteins that stimulate viral resistance in cells.

Intergenic regions: DNA sequences located between genes that comprise a large percentage of the human genome with no known function.

Introgression: Backcrossing of hybrids of two plant populations to introduce new genes into a wild population.

Intron: A noncoding DNA sequence within a gene that is initially transcribed into messenger RNA but is later snipped out. See Coding, DNA, Messenger RNA, Transcription.

Invasiveness: Ability of a plant to spread beyond its introduction site and become established in new locations where it may provide a deliterious effect on organisms already existing there.

Ion: A charged particle.

Isotope: One of two or more forms of an element that have the same number of protons (atomic number) but differing numbers of neutrons (mass numbers). Radioactive isotopes are commonly used to make DNA probes and metabolic tracers.

Joining (J) segment: A small DNA segment that links genes to yield a functional gene encoding an immunogobulin.

Jupiter-Brain: A posthuman being of extremely high computational power and size. This is the archetypal concentrated intelligence. The term originated due to an idea by Keith Henson that nanomachines could be used to turn the mass of Jupiter into computers running an upgraded version of himself.

Kanamycin: An antibiotic of the aminoglycoside family that poisons translation by binding to the ribosomes.

Karyotype: All of the chromosomes in a cell or an individual organism, visible through a microsope during cell division.

Kb (kilobase): The number of base pairs (in thousands) that denote the size of a nucleic acid fragment.

Kilojoule: Standard unit of energy equal to 1000 joules, or 0.24 kilocalories.

Lag phase: The initial growth phase, during which cell number remains relatively constant prior to rapid growth. See growth phase.

Lane: On a gel or a blot, a lane belongs to one sample. It contains fragments of different sizes that were separated by gel electrophoresis. See band, blot.

Linker: A short, double-stranded oligonucleotide containing a restriction endonuclease recognition site, which is ligated to the ends of a DNA fragment.

Liposome: A type of nanoparticle made of lipids, or fat molecules, surrounding a water core. Liposomes, several of which are widely used to treat infectious diseases and cancer, were the first type of nanoparticle to be used to create therapeutic agents with novel characteristics. Membrane-bound vesicles constructed in the laboratory to transport biological molecules.

Liquid Crystal Display: LCD is the predominant technology used in flat panel displays. The principle that makes the display work is this: A crystalís alignment can be altered with an electric current. If the crystal is lined up one way ñ it will allow the light waves to pass through a polarized filter, but if the electric current alters the crystalís alignment, it will guide light so that the polarized filter blocks the light. By densely packing red, blue and green light emitting crystals next to each other on a sheet (ìcalled a substrateî), one can create a full color display. The great thing about LCD is that the crystals can be packed together closely, allowing for a higher-resolution, finer-detail display. The con is that LCDs are somewhat fragile, require a lot of power and are relatively less bright. Liquid Crystal Display — display device employing light source and electrically alterable optically active thin film.

Locus (plural = loci): A specific location or site on a chromosome.

Logarithmic phase (log or exponential growth phase): The steepest slope of the growth curve—the phase of vigorous growth during which cell number doubles every 20-30 minutes.

Ligase (DNA ligase): An enzyme that catalyzes a condensation reaction that links two DNA molecules via the formation of a phosphodiester bond between the 3' hydroxyl and 5' phosphate of adjacent nucleotides.

Ligation: Joining two fragments of DNA end to end. See sticky ends.

Light Emitting Diodes (LEDs): Light Emitting Diodes work on a completely different concept. Traditionally LEDs are created from two semiconductors. By running current in one direction across the semiconductor the LED emits light of a particular frequency (hence a particular color) depending on the physical characteristics of the semiconductor used. The semiconductor is covered with a piece of plastic that focuses the light and increases the brightness. These semiconductors are very durable, there is no filament, they donít require much power, theyíre brighter and they last a long time. By densely packing red, blue and green LEDs next to each other on a substrate one can create a display.

Lineage: A chart that traces the flow of genetic information from generation to generation.

Linkage: The frequency of coinheritance of a pair of genes and/or genetic markers, which provides a measure of their physical proximity to one another on a chromosome. Measures the physical distance between two genes. Genes that are close together are unlikely to segregate in a sexual cross. Distant genes segregate independently are are then said to be unlinked.

Linked genes/markers: Genes and/or markers that are so closely associated on the chromosome that they are coinherited in 80% or more of cases.

Lawn: A uniform and uninterrupted laver of bacterial growth, in which individual colonies cannot be observed.

Lectin: Protein that binds tightly to a specific sugar. Abundant lectins derived from plant seeds are often used as affinity reagents to purify glycoproteins or to detect them on the surface of cells.

Legume: A member of the pea family that possesses root nodules containing nitrogen-fixing bacteria.

Library, DNA: A large set of clones of DNA fragments from a particular organism. DNA libraries are usually maintained in *E. coli* or in bacteriophages.

Life (lifetime): The length of time the sensor can be used before its performance changes.

Life Science Applications of Deep Red T2-MP EviTag Quantum Dots: The new T2-MP EviTags are available in the deep red wavelengths used in life science research – Macoun Red (660nm) and Jonamac Red (680nm) – with carboxyl, amine, biotin and non-functionalized surfaces. Having a biologically compatible quantum dot label like the T2-MP EviTags allows further exploration and development of in vivo assays, following cells inside living organism to get a new understanding of life. In particular, it offers the possibility to create new classes of assays for use where current testing materials could produce cytotoxicity or cell death. Potential applications include live cell and even whole animal imaging, blood cancer assays, and numerous other applications.

Ligand: In protein chemistry, a small molecule that is (or can be) bound by a larger molecule is termed a ligand. In organometallic chemistry, a moiety bonded to a central metal atom is also termed a ligand; the latter definition is more common in general chemistry.

Lysogen: A bacterial cell whose chromosome contains integrated viral DNA.

Lysogenic: A type or phase of the virus life cycle during which the virus integrates into the host chromosome of the infected cell, often remaining essentially dormant for some period of time. See Lysogen.

Lytic: A phase of the virus life cycle during which the virus replicates within the host cell, releasing a new generation of viruses when the infected cell lyses.

Machine-phase chemistry: The chemistry of systems in which all potentially reactive moieties follow controlled trajectories (e.g., guided by molecular machines working in vacuum).

Macroscale: Larger than nanoscale; often implies a design that humans can directly interact with; too large to be built by a single assembler (one cubic micron of diamond contains 176 billion atoms).

Magnetization (M): The total magnetic moment per unit volume of material. Also, a measure of the contribution to the magnetic flux by some material within an H field. The magnitude of M is proportional to the applied field as: M = ÷m × H, with ÷m the magnetic susceptibility.

Magnetostrictive material: A material that changes dimension in the presence of a magnetic field or generates a magnetic field when mechanically deformed.

Malignant: Having the properties of cancerous growth.

Manufacturing: The ability to make products, in this case ranging from clothing, to electronics, to medical devices, to books, to building materials, and much more.

Meat Machine: AKA Cabinet Beast. A box containing

assemblers and raw material, within which is formed meat [or whatever else it was programmed to make].

Mechanical: Pertaining to the positions and motions of atoms, as defined by the positions of their nuclei; see electronic. A purely mechanical device can be described in terms of atomic positions and motions without reference to electronic properties, save through their effect on the potential energy function.

Mechanochemistry: Chemistry accomplished by mechanical systems directly controlling the reactant molecules; the formation or breaking of chemical bonds under direct mechanical control. In this volume, the chemistry of processes in which mechanical systems operating with atomic-scale precision either guide, drive, or are driven by chemical transformations. In general usage, the chemistry of processes in which energy is converted from mechanical to chemical form, or vice versa.

Mechanosynthesis: Chemical synthesis controlled by mechanical systems operating with atomic-scale precision, enabling direct positional selection of reaction sites; synthetic applications of mechanochemistry. Suitable mechanical systems include AFM mechanisms, molecular manipulators, and molecular mill systems. Processes that fall outside the intended scope of this definition include reactions guided by the incorporation of reactive moieties into a shared covalent framework (i.e., conventional intramolecular reactions), or by the binding of reagents to enzymes or enzymelike catalysts. Molecular tools with chemically specific tip structures can be used, sequentially, to modify a work piece and build a wide range of molecular structures.

Mechatronics: The study of the melding of AI and

electromechanical machines to make machines that are greater than the sum of their parts. The synergistic combination of precision mechanical engineering with electronic control.

Megabase cloning: The cloning of very large DNA fragments.

Megabyte (MB): 220 (= 1,048,576, or about one million) bytes of information.

Megahertz (Mhz): One million cycles per second

Meiosis: The reduction division process by which haploid gametes and spores are formed, consisting of a single duplication of the genetic material followed by two mitotic divisions.

Meme: An idea that replicates through a society as it is propagated through person-to-person interaction, both direct and indirect. Memetics is a field of study that focuses on memes' role in the evolution of a culture. Self-reproducing idea or other information pattern which is propagated in ways similar to that of a gene. [Richard Dawkins, 1976]

Messenger RNA (mRNA): The class of RNA molecules that copies the genetic information from DNA, in the nucleus, and carries it to ribosomes, in the cytoplasm, where it is translated into protein.

Metabolism: The biochemical processes that sustain a living cell or organism.

Metallothionein: A protective protein that binds heavy metals, such as cadmium and lead.

Micellar nanocontainers: Block copolymer micelles are water- soluble biocompatible nanocontainers with great potential for delivering hydrophobic drugs. An understanding of their cellular distribution is essential to achieving selective delivery of drugs at the subcellular level. R. Savic, L. Luo, A. Eisenberg, D. Maysinger, Micellar nanocontainers distribute to defined cytoplasmic organelles Science 300 (5619): 615- 618, Apr. 25, 2003

Micelles: Micelles are small, spherical structures composed of molecules that attract one another to reduce surface tension. The head of the molecule is hydrophilic, meaning it likes water, while the interior portion is hydrophobic, meaning it avoids water.

Microbial mats (biofilms): Layered groups or communities of microbial populations.

Microelectronics: Narrower terms: MEMS, nanoelectronics, optoelectronics, SED Single Electron Devices. Related terms: molecular electronics, semiconductors

Microencapsulation: Individually encapsulated small particles.

Microinjection: A means to introduce a solution of DNA, protein, or other soluble material into a cell using a fine microcapillary pipet.

Microspectrophotometry: Analytical technique for studying substances present at enzyme concentrations in single cells, in situ, by measuring light absorption. Light from a tungsten strip lamp or xenon arc dispersed by a grating monochromator illuminates the optical system of a microscope. The absorbance of light is measured (in nanometers) by comparing the difference between the image of the sample and a reference image. MeSH 1990

Microspheres: Microspheres are generally defined as small spheres made of any material and sized from about 0.5 μm to 100 μm. Similar, but smaller spheres sized 10 to 500 nm are called *nanospheres*. Ideally, microspheres are completely spherical and homogeneous in size, although less perfect particles are often termed microspheres as well. Depending on the preparation method and material used, microspheres show a typical size distribution which often deviates from the mono-sized ideal.

Microstructure: In material engineering the structural features of a material such as grain boundaries, grain size and structure, subject to observation under a microscope, selective etching etc. In MEMS microstructure unfortunately is also used to designate a micromachined feature. The last decade has seen rapid developments in the fabrication, characterization and conceptual understanding of synthetic microstructures in many different material systems including silicon, III-V and II-VI semiconductors, metals, ceramics and organics. The objective of this journal [Superlattices and Microstructures] is to provide a common interdisciplinary platform for the publication of the latest research results on all such "nanostructures" with dimensions in the range of 1-100 nm; the unifying theme here being the dimensions of these artificial structures rather than the material system in which they are fabricated.

Microsystem: A microscale machine that can sense information from the environment and act accordingly. Outside the U.S., it can also refer to microelectromechanical systems (MEMS). [smalltimes glossary, 2002]

MicroTas, micro Total Analysis Systems, uTAS: Although initial research dates back to the early 1970's, the field of micro- TAS formally started in 1990, when Manz et al described the possibility of creating microsystems that would take care of many or all the traditional analytical steps involved in a biochemical analysis (sample introduction, handling, extraction, purification, concentration, filtration, analysis, detection) .... Micro- TAS offer many advantages over traditional analysis systems. Low power consumption and small reaction volumes, faster analysis, ultrasensitive detection, and minimal human intervention are key parameters in the development of micro- TAS. Most biochemical reactions take place in liquid environments. Hence, the development of MicroTAS is intrinsically linked to the design of liquid handling micro- devices. [Biomedical Applications Group (GAB) Centro Nacional de Microelectronica (CNM- IMB) Bellaterra, Spain, 2000]

Microtubule: A hollow cylindrical tube used for the transport of materials inside cells.

Miller indices: A set of 3 integers (4 for hexagonal) that designate crystallographic planes, as determined from reciprocals of fractional axial intercepts.

Millimeter: One thousandth of a meter, or about 1/26 of an inch.

Mitosis: The replication of a cell to form two daughter cells with identical sets of chromosomes.

Molecular - Nanotechnology: The technology of precisely-constructed molecular-scale machines; from nanometer: a billionth of a meter. For a good introduction, see What is Nanotechnology? or Ralph C. Merkle's nanotechnology page.

Molecular assembler: A general-purpose device for molecular manufacturing, able to guide chemical reactions by positioning individual molecules to atomic accuracy (e.g. mechanosynthesis) and to construct a wide range of useful and stable molecular structures according to precise specifications. Also known as an assembler, a molecular assembler is a molecular machine that can build a molecular structure from its component building blocks.

Molecular Beam Epitaxy: [MBE] Process used to make compound (multi-layer) semiconductors. Consists of depositing alternating layers of materials, layer by layer, one type after another (such as the semiconductors gallium arsenide and aluminum gallium arsenide).

Molecular biology: The study of the biochemical and molecular interactions within living cells.

Molecular cloning: The biological amplification of a specific DNA sequence through mitotic division of a host cell into which it has been transformed or transfected.

Molecular genetics: The study of the flow and regulation of genetic information between DNA, RNA, and protein molecules.

Molecular markers: Genetic traits which are detectable on gels or blots and can be used to construct genetic maps. Molecular markers usually have no known function. RFLPs are molecular markers.

Molecular mechanics models: Many of the properties of molecular systems are determined by the molecular potential energy function. Molecular mechanics models approximate this function as a sum of 2-atom, 3-atom, and 4-atom terms, each determined by the geometries and bonds of the component atoms. The 2-atom and 3-atom terms describing bonded interactions roughly correspond to linear springs.

Molecular medicine: A variety of pharmaceutical techniques and gene therapies that address specific molecular diseases or molecular defects in biological systems. A variety of pharmaceutical techniques and therapies in use today. Studying molecules as they relate to health and disease, and manipulating those molecules to improve the diagnosis, prevention, and treatment of disease.

Molecular mill: A mechanochemical processing system characterized by limited motions and repetitive operations without programmable flexibility.

Molecular motors: Protein based machines that are involved in or cause movement such as the rotary devices (flagellar motor and the F1 ATPase) or the devices whose movement is directed along cytoskeletal filaments (myosin, kinesin and dynein motor families). [MeSH, 1999]

Molecular Nanoscience: An emerging interdisciplinary field that combines the study of molecular/ biomolecular systems with the science and technology of nanoscale structures and systems. The potential applications for this research are very broad and include such possibilities as 1) the use of biomolecules and cellular systems to self- assemble nanoelectronic circuitry and other nanoscale structures and 2) the use of lamellar host frameworks containing nano pores that can be tailored to include guest molecules for separation of chemicals for pharmaceutical and other applications. Molecular Nanoscience Alliance for Interdisciplinary Studies and Activities, Univ. of Minnesota, US,

Molecular Wire: A molecular wire—the simplest electronic component—is a quasi-one-dimensional molecule that can transport charge carriers (electrons or holes) between its ends. [Michael D Ward]

Molecularly imprinted polymers MIPs: A new class of

materials that have artificially created receptor structures. Since their discovery in 1972, MIPs have attracted considerable interest from scientists and 3engineers involved with the development of chromatographic absorbents, membranes, sensors and enzyme and receptor mimics.

Monoclonal antibodies: Immunoglobulin molecules of single- epitope specificity that are secreted by a clone of B cells.

Monoculture: The agricultural practice of cultivating crops consisting of genctically similar organisms.

Monogenic: Controlled by or associated with a single gene.

Multi-locus probe: A probe that hybridizes to a number of different sites in the genome of an organism.

Mutagen: Any agent or process that can cause mutations. See Mutation.

Mutation: An alteration in DNA structure or sequence of a gene. A spontaneous or induced genetic change in the DNA of an organism. Most mutations are undesirable, but a few are useful such as dwarfing genes in cereal crops that allow the plants to stand better.

Mycorrhizae: Fungi that form symbiotic relationships with roots of more developed plants.

Nanobiology: Many fundamental biological functions are carried out by molecular machineries that have the sizes of 1-100 nm. You find many examples in molecular biology and cell biology: single enzymes, transcription complex, ribosome, transport complex, nuclear pore, and so on. To understand the functions of these machineries, one has to describe their movements, changes in their shapes, and their localization. This means the mechanistic study is equivalent to dynamic morphology at this level of size, making a new field that merges mechanistic biology and morphology. The emergence of nanobiology depended on the invention of nano- technology: scanning probe microscopy, modern optical techniques, and micro- manipulating techniques. This concept of nanobiology was first proposed in a group study named "Biological Nano- Mechanisms", which was supported by Japanese Agency of Science and Technology (1992-1997). [National Institute of Genetics, Japan]

Nanobiotechnology: An emerging area of scientific and technological opportunity. Nanobiotechnology applies the tools and processes of nano/ microfabrication to build devices for studying biosystems. Researchers also learn from biology how to create better micro-nanoscale devices. The Nanobiotechnology Center (NBTC), a National Science Foundation, Science and Technology Center is characterized by its highly interdisciplinary nature and features a close collaboration between life scientists, physical scientists, and engineers. [NBTC, NanoBiotechnology Center, Cornell Univ. US, 2002] Use of nanotechnology[ies] in the life sciences. These may include, but are not limited to, therapeutics, medical devices/implants, biosensors, and tools for the development of drugs. Applying the tools and processes of MNT to build devices for studying biosystems, in order to learn from biology how to create better nanoscale devices. Should hasten the creation of useful micro devices that mimic living biological systems.

Nanocentrifuge: In medical nanorobotics, a proposed nanodevice that can spin materials at very high speed, imparting rotational accelerations of up to one trillion gravities (g's), thus permitting rapid sortation.

Nanoscience: Nanoscience is the science of nanoscale phenomena. This is a very complicated area, since Quantum Mechanics is often needed to describe the physical properties of atoms and molecules properly. What makes it even more complicated is that nanostructured materials can still be quite a challenge to describe Quantum Mechanically, since the equations are feindishly difficult—this is where supercomputers come in to make the job approachable. The scientific discipline seeking to increase our knowledge and understanding of nanoscale phenomena, i.e. science on the scale of 0.1 nm to 100 nm.

Nanosensor: A chemical or physical sensor constructed using nanoscale components, usually microscopic or submicroscopic in size.

Nanosources: sources that emit light from nanometre-scale volumes.

National Institutions of Health (NIH): A nonregulatory agency which has oversight of research activities that the agency funds.

National Science Foundation (NSF): A nonregulatory agency which has oversight of biotechnology research activities that the agency funds.

Natural selection: The differential survival and reproduction of organisms with genetic characteristics that enable them to better utilize environmental resources.

Neophile: One who welcomes the future and who enjoys change and evolution.

Nick translation: A procedure for making a DNA probe in which a DNA fragment is treated with DNase to produce single-stranded nicks, followed by incorporation of radioactive nucleotides from the nicked sites by DNA polymerase I.

Nicked circle (relaxed circle): During extraction of plasmid DNA from the bacterial cell, one strand of the DNA becomes nicked. This relaxes the torsional strain needed to maintain supercoiling, producing the familiar form of plasmid.

Nitrocellulose: A membrane used to immobilize DNA, RNA, or protein, which can then be probed with a labeled sequence or antibody.

Nitrogen fixation: The conversion of atmospheric nitrogen to biologically usable nitrates.

Nitrogenous bases: The purines (adenine and guanine) and pyrimidines (thymine, cytosine, and uracil) that comprise DNA and RNA molecules.

Nodule: The enlargement or swelling on roots of nitrogen-fixing plants. The nodules contain symbiotic nitrogen-fixing bacteria. See Nitrogen fixation.

Nontarget organism: An organism which is affected by an interaction for which it was not the intended recipient.

Northern hybridization: (Northern blotting). A procedure in which RNA fragments are transferred from an agarose gel to a nitrocellulose filter, where the RNA is then hybridized to a radioactive probe.

Novel foods: Novel foods are products that have never been used as a food; foods which result from a process that has not previously been used for food; or, foods that have been modified by genetic manipulation. This last category of foods have been described as genetically modified foods (often referred to as GM foods, genetically engineered foods or biotechnology-derived foods). *Definition from Health Canada Web site*

Nuclease: Any enzyme that cuts nucleic acids. See restriction enzyme. A class of enzymes that degrades DNA and/or RNA molecules by cleaving the phosphodiester bonds that link adjacent nucleotides. In deoxyribonuclease (DNase), the substrate is DNA. In endonuclease, it cleaves at internal sites in the substrate molecule. Exonuclease progressively cleaves from the end of the substrate molecule. In ribonuclease (RNase), the substrate is RNA. In the S1 nuclease, the substrate is single-stranded DNA or RNA.

Nucleic acids: The two nucleic acids, deoxyribonucleic acid (DNA) and ribonucleic acid (RNA), are made up of long chains of molecules called nucleotides. See DNA, RNA, Nucleotides.

Nuclein: The term used by Friedrich Miescher to describe the nuclear material he discovered in 1869, which today is known as DNA.

Nucleoside analog: A synthetic molecule that resembles a naturally occuring nucleoside, but that lacks a bond site needed to link it to an adjacent nucleotide.

Nucleotide: A building block of DNA and RNA, consisting of a nitrogenous base, a five-carbon sugar, and a phosphate group. Together, the nucleotides form codons, which when strung together form genes, which in turn link to form chromosomes.

Nucleus: The membrane-bound region of a eukaryotic cell that contains the chromosomes. A cellular organelle in plants and animals that contains the chromosomes which in turn are composed of DNA plus protein.

Occupational Safety and Health Administration (OSHA): One of the U.S. agencies responsible for regulation of biotechnology. The major law under which the agency has regulatory powers is the Occupational Safety and Health Act.

Oligonucleotide: A DNA polymer composed of only a few nucleotides.

Oncogene: A gene that contributes to cancer formation when mutated or inappropriately expressed.

Oncogenesis: The progression of cytological, genetic, and cellular changes that culminate in a malignant tumor.

Open pollination: Pollination by wind, insects, or other natural mechanisms.

Open reading frame: A long DNA sequence that is uninterrupted by a stop codon and encodes part or all of a protein. (

Operator: A prokaryotic regulatory element that interacts with a repressor to control the transcription of adjacent structural genes.

Orbital: In the approximation that each electron in a molecule has a distinct, independent wave function, the spatial distribution of an electron wave function corresponds to a molecular orbital. These, in turn, can be approximated as sums of contributions from the orbitals characteristic of the isolated atoms. An electron added to a molecule—or, similarly, one excited to a higher-energy state within a molecule—would occupy a state with a different wave function from the rest; an unoccupied state of this kind corresponds to an unoccupied molecular orbital. Orbital-symmetry effects on reaction rates arise when a reaction requires overlap between two lobes of the orbitals on each of two reagents: if the algebraic signs of the wave functions in the facing lobes do not match, bond formation between those orbitals is prohibited.

Organelle: A cell structure that carries out a specialized function in the life of a cell.

Origin of replication: The nucleotide sequence at which DNA synthesis is initiated.

Overlapping reading frames: Start codons in different reading frames generate different polypeptides from the same DNA sequence.

Ovum: A female gamete.

Paleontology: The study of the fossil record of past geological periods and of the phylogenetic relationships between ancient and contemporary plant and animal species.

Palindromic sequence: A DNA locus whose 5'-to-3' sequence is identical on each DNA strand. The sequence is the same when one strand is read left to right and the other strand is read right to left. Recognition sites of many restriction enzymes are palindromic. See DNA.

pAMP: Ampicillin-resistant plasmid developed for this laboratory course.

Parasitism: The closee association of two or more dissimilar organisms where the association is harmful to at least one. See Commensalism, Parasitism, Symbiosis.

Particle bombardment: Using metal particles to blast DNA into cells for the purpose of genetic engineering. A gene gun is used to propel the metal particles.

Pathogen: Organism which can cause disease in another organism.

pBR322: A derivation of ColE1, one of the first plasmid vectors widely used.

PCR: Polymerase chain reaction. An amazingly sensitive technique that allows the specific amplification of extremely small amounts of particular DNA fragments using DNA polymerase and specific primers.

Pedigree: A diagram mapping the genetic history of a particular family.

Percent Accepted Mutation: A unit introduced by Dayhoff et al. to quantify the amount of evolutionary change in a protein sequence. 1.0 PAM unit, is the amount of evolution which will change, on average, 1% of amino acids in a protein sequence. A PAM(x) substitution matrix is a look-up table in which scores for each amino acid substitution have been calculated based on the frequency of that substitution in closely related proteins that have experienced a certain amount (x) of evolutionary divergence.

Persistence: Ability of an organism to remain in a particular setting for a period of time after it is introduced.

Pesticide: A substance that kills harmful organisms (for example, an insecticide or fungicide).

Phenotype: The observable characteristics of an organism, the expression of gene alleles (genotype) as an observable physical or biochemical trait. See Genotype.

Pheromone: A hormone-like substance that is secreted into the environment.

Phosphatase: An enzyme that hydrolyzes esters of phosphoric acid, removing a phosphate group.

Phosphodiester bond: A bond in which a phosphate group joins adjacent carbons through ester linkages. A condensation reaction between adjacent nucleotides results in a phosphodiester bond between 3' and 5' carbons in DNA and RNA.

Phospholipid: A class of lipid molecules in which a phosphate group is linked to glycerol and two fatty acyl groups. A chief component of biological membranes.

Phosphorylation: The addition of a phosphate group to a compound.

Physical map: A map showing physical locations on a DNA molecule, such as restriction sites, and sequence-tagged sites.

Piezoelectric material: A ferroelectric material in which an electrical potential difference is created due to mechanical deformation, or conversely, in which the application of a voltage causes dimensional changes in the material.

Piezoelectric: A piezoelectric material expands or contracts when a voltage is applied across it. Piezoelectric tubes are often used to move the tip in an AFM or STM. If a piezoelectric crystal is compressed, a voltage appears across it. This effect is sometimes used to measure displacements. A commonly used piezoelectric is PZT (Pb,Zr)TiO3.

Plaque: A clear spot on a lawn of bacteria or cultured cells where cells have been Iysed by viral infection.

Plasmid (p): A circular DNA molecule, capable of autonomous replication, which typically carries one or more genes encoding antibiotic resistance proteins. Plasmids can transfer genes between bacteria and are important tools of transformation for genetic engineers.

Pleiotrophy: The effect of a particular gene on several different traits.

Point mutation: A change in a single base pair of a DNA sequence in a gene.

Poly(A) polymerase: Catalyzes the addition of adenine residues to the 3' end of pre-mRNAs to form the poly(A) tail.

Polyacrylamide gel electrophoresis: Electrophoresis through a matrix composed of a synthetic polymer, used to separate proteins, small DNA, or RNA molecules of up to 1000 nucleotides. Used in DNA sequencing.

Polyclonal antibodies: A mixture of immunoglobulin molecules secreted against a specific antigen, each recognizing a different epitope.

Polygenic: Controlled by or associated with more than one gene.

Polylinker: A short DNA sequence containing several restriction enzyme recognition sites that is contained in cloning vectors.

Polymer: A molecule composed of repeated subunits.

Polymerase (DNA): Synthesizes a double-stranded DNA molecule using a primer and DNA as a template.

Polymerase chain reaction (PCR): A procedure that enzymatically amplifies a DNA polymerase.

Polymerase: Any enzyme that adds subunits to chains of macromolecules. Example is DNA polymerase.

Polymorphisms: Variant forms of a particular gene that occur simultaneously in a population.

Polynucleotide: A DNA polymer composed of multiple nucleotides.

Polypeptide (protein): A polymer composed of multiple amino acid units linked by peptide bonds.

Polyploid: A multiple of the haploid chromosome number that results from chromosome replication without nuclear division.

Polysaccharide: A polymer composed of multiple units of monosaccharide (simple sugar).

Polyvalent vaccine: A recombinant organism into which has been cloned antigenic determinants from a number of different disease-causing organisms.

Population: A local group of organisms belonging to the same species and capable of interbreeding.

Primary cell: A cell or cell line taken directly from a living organism, which is not immortalized.

Primer: A short DNA or RNA fragment annealed to single-stranded DNA, from which DNA polymerase extends a new DNA strand to produce a duplex molecule.

Prion: See Proteinaceous infectious particle.

Probe: A single-stranded DNA that has been radioactively labeled and is used to identify complementary sequences in genes or DNA fragments of interest. A fragment of DNA, usually labeled with radioactive 32P, that detects homologous sequences on southern blots.

Prokaryote: A bacterial cell lacking a true nucleus; its DNA is usually in one long strand. See Eukaryote.

Promoter: A region of DNA extending 150-300 bp upstream from the transcription start site that contains binding sites for RNA polymerase and a number of proteins that regulate the rate of transcription of the adjacent gene.

Protein: Molecules composed of amino acids. Proteins constitute the enzymes and many of the structural components of cells. A polymer of amino acids linked via peptide bonds and which may be composed of two or more polypeptide chains. Proteins do the work in cells. They can be part of structures (such as cell walls, organelles, etc), regulate reactions that take place in the cell or they can serve as enzymes, which speed-up reactions. Everything you see in an organism is either made of proteins or the result of a protein action.

Quantum Mechanics: A largely computational physical theory explaining the behavior of quantum phenomena, which incorporates the theory of special relativity. Despite dilignet attempts, general relativity has not been sucessfully incorporated into quantum mechanics. A physical model of chemical and optical phenomena, as well as the behaviour of matter in general on a small scale. Quantum mechanics describes a system of particles in terms of a wave function defined over the configuration space of the system. Although the concept of particles having distinct locations is implicit in the potential energy function that determines the wave function (e.g., of a ground-state system), the observable dynamics of the system cannot be described in terms of the motion of such particles from point to point. In describing the energies, distributions, and behaviors of electrons in nanometer-scale structures, quantum mechanical methods are necessary. Electron wave functions help determine the potential energy surface of a molecular system, which in turn is the basis for classical descriptions of molecular motion. Nanomechanical systems can almost always be described in terms of classical mechanics, with occasional quantum mechanical corrections applied within the framework of a classical model.

rDNA: Def. 1. Ribosomal DNA; DNA which codes for ribosomes

Reading frame: A series of triplet codons beginning from a specific nucleotide. Depending on where one begins, each DNA strand contains three different reading frames.

Recessive gene: Characterized as having a phenotype expressed only when both copies of the gene are mutated or missing.

Recessive(-acting) oncogene, (anti-oncogene): A single copy of this gene is sufficient to suppress cell proliferation; the loss of both copies of the gene contributes to cancer formation.

Recessive: Refers to the member of a pair of alleles that fails to be expressed in the phenotype of the organism when the dominant member is present. Also refers to the phenotype of an individual that has only the recessive allele.

Recognition sequence (site): A nucleotide sequence—composed typically of 4, 6, or 8 nucleotides—that is recognized by a restriction endonuclease. Type II enzyrnes cut (and their corresponding modification enzymes methylate) within or very near the recognition sequence.

Recombinant DNA: The process of cutting and recombining DNA fragments from different sources as a means to isolate genes or to alter their structure and function.

Recombination frequency: The frequency at which crossing over occurs between two chromosomal loci—the probability that two loci will become unlinked during meiosis.

Refuge: A refuge is any host plant (non-Bt crop, potatoes, oats, sorghum, and some weeds) not producing Bt proteins or not being treated with conventional Bt formulations. The purpose of the refuge is to supply a source of susceptible insects that may mate with resistant insects of the same species emerging from nearby Bt crop fields. If the refuge is large enough, susceptible insect numbers should outnumber resistant individuals, and in theory, a greater number of the next generation offspring will be susceptible to the Bt crop.

Regulatory gene: A gene whose protein controls the activity of other genes or metabolic pathways.

Relaxed plasmid: A plasmid that replicates independently of the main bacterial chromosome and is present in 10-500 copies per cell.

Renature: The reannealing (hydrogen bonding) of single-stranded DNA and/or RNA to form a duplex molecule.

Replicon: A chromosomal region containing the DNA sequences necessary to initiate DNA replication processes.

Repressor: A DNA-binding protein in prokaryotes that blocks gene transcription by binding to the operator.

Resistance: The evolved ability of a species to survive during a stressful event or set of circumstances.

Restriction endonuclease (enzyme): A class of endonucleases that cleaves DNA after recognizing a specific sequence, such as BamH1 (GGATCC), EcoRI (GAATTC), and HindIII (AAGCTT). Type I. Cuts nonspecifically a distance greater than 1000 bp from its recognition sequence and contains both restriction and methylation activities. Type II. Cuts at or near a short, and often symmetrical, recognition sequence. A separate enzyme methylates the same recognition sequence. Type III. Cuts 24-26 bp downstream from a short, asymmetrical recognition sequence. Requires ATP and contains both restriction and methylation activities.

Restriction enzyme: A class of enzymes that cut DNA at specific sequences called restriction sites. Restriction enzymes were key to making genetic engineering possible. See RFLP.

Restriction-fragment-length polymorphism (RFLP): Differences in nucleotide sequence between alleles at a chromosomal locus result in restriction fragments of varying lengths detected by Southern analysis.

Retrovirus: A member of a class of RNA viruses that utilizes the enzyme reverse transcriptase to reverse copy its genome into a DNA intermediate, which integrates into the hostcell chromosome. Many naturally occurring cancers of vertebrate animals are caused by retroviruses.

Reverse genetics: Using linkage analysis and polymorphic markers to isolate a disease gene in the absence of a known metabolic defect, then using the DNA sequence of the cloned gene to predict the amino acid sequence of its encoded protein.

Reverse transcriptase (RNA-dependent DNA polymerase): An enzyme isolated from retrovirus-infected cells that synthesizes a complementary (c)DNA strand from an RNA template.

RFLP: Restriction fragment length polymorphism. RFLPs are generated by digesting DNA with restriction enzymes. The DNA is separated by size by gel electrophoresis. Slight differences in homologous fragments may exist between individuals. These length polymorphisms are RFLPs. See blot.

Rhizobia: Bacteria in a symbiotic relationship with leguminous plants that results in nitrogen fixation. See Nitrogen fixation.

Rhizosphere: The soils region on and around plant roots.

Ribosomal RNA (rRNA): The RNA component of the ribosome.

Ribosome: Cellular organelle that is the site of protein synthesis during translation. See Organelle, Translation.

Ribosome-binding site: The region of an mRNA molecule that binds the ribosome to initiate translation.

Risk assessment: Defined as a formalized basis for the objective evaluation of risk in a manner in which assumptions and uncertainties are clearly considered and presented.

RNA (ribonucleic acid): An organic acid composed of repeating nucleotide units of adenine, guanine, cytosine, and uracil, whose ribose components are linked by phosphodiester bonds.

RNA polymerase: Transcribes RNA from a DNA template.

Satellite RNA (viroids): A small, self-splicing RNA molecule that accompanies several plant viruses, including tobacco ringspot virus.

Scanning Near Field Optical Microscopy: A method for observing local optical properties of a surface that can be smaller than the wavelength of the light used. A type of scanning probe microscopy that maps the near-field optical properties of an interface.

Scanning Probe Microscopy (SPM): A method for imaging nanoscale features of surfaces by scanning a sensor (probe) over a surface. Near-field effects such as tunneling, van der Waals forces, local fields and more are serially detected at localized points on the surface and used to create an SPM image.

Scanning Thermal Microscopy (SThM): A type of scanning probe microscopy that maps the local temperature and thermal conductivity of an interface. A method for observing local temperatures and temperature gradients on a surface. [NTN]

Scanning Tunneling Microscope (STM): An instrument able to image conducting surfaces to atomic accuracy; has been used to pin: molecules to a surface. A device in which a sharp conductive tip is moved across a conductive surface close enough to permit a substantial tunneling current (typically a nanometer or less). In a common mode of operation, the voltage is kept constant and the current is monitored and kept constant by controlling the height of the tip above the surface; the result, under favorable conditions, is an atomic-resolution map of the surface reflecting a combination of topography and electronic properties. The STM has been used to manipulate atoms and molecules on surfaces. A type of scanning probe microscopy that maps the local electrical properties of an interface.

Schwann cell: Glial cell responsible for forming myelin sheaths in the peripheral nervous system.

Science: The process of developing a systematized knowledge of the world through the variation and testing of hypotheses. The pursuit of knowledge and understanding, from the Latin term *scientia*, which means 'knowledge'.

Selectable marker genes: Genetic engineers must be able to select cells that have received the transgene from those that have not. Selecting out transgenic cells is done by co-transforming the cells with the transgene plus an additional gene called a selectable marker gene. Selectable marker genes are genes that encode easily detectable traits making transgenic cells discernible from non-transgenic cells. The two most commonly used selectable marker genes encode the traits of herbicide and antibiotic resistance.

Selectable marker: A gene whose expression allows one to identify cells that have been transforrned or transfected with a vector containing the marker gene.

Self-pollination: Pollen of one plant is transferred to the female part of the same plant or another plant with the same genetic makeup.

Semiconservative replication: During DNA duplication, each strand of a parent DNA molecule is a template for the synthesis of its new complementary strand. Thus, one half of a preexisting DNA molecule is conserved during each round of replication.

Sequence hypothesis: Francis Crick's seminal concept that genetic information exists as a linear DNA code; DNA and protein sequence are colinear.

Sequence: Noun: the particular order of nucleotides in a DNA or RNA fragment. Verb: to determine the particular order of nucleotides in a strand of DNA.

Sequence-tagged site (STS): A unique (single-copy) DNA sequence used as a mapping landmark on a chromosome.

Sexual reproduction: The process where two cells (gametes) fuse to form one hybrid, fertilized cell. See Asexual reproduction, Gamete, Hybrid.

Signal transduction: The biochemical events that conduct the signal of a hormone or growth factor from the cell exterior, through the cell membrane, and into the cytoplasm. This involves a number of molecules, including receptors, pro- teins, and messengers.

Site-directed mutagenesis: The process of introducing spe- cific base-pair mutations into a gene.

Small nuclear RNA (snRNA): Short RNA transcripts of 100-300 bp that associate with proteins to form small nuclear ribonucleoprotein particles (snRNPs), which participate in RNA processing.

Somatic cell gene therapy: The repair or replacement of

a defective gene within somatic tissue.

Somatic cell: Any nongerm cell that composes the body of an organism and which possesses a set of multiploid chromosomes (diploid in most organisms).

Southern hybridization (Southern blotting): A procedure in which DNA restriction fragments are transferred from an agarose gel to a nitrocellulose filter, where the denatured DNA is then hybridized to a radioactive probe (blotting).

Species: A classification of related organisms that can freely interbreed.

Spore: A form taken by certain microbes that enables them to exist in a dormant stage. It is an asexual reproductive cell. See Asexual reproduction, Dormant.

Stationary phase: The plateau of the growth curve after log growth, during which cell number remains constant. New cells are produced at the same rate as older cells die.

Sticky end: A protruding, single-stranded nucleotide sequence produced when a restriction endonuclease cleaves off center in its recognition sequence. After cutting with restriction enzymes, the ends of DNA fragments are "sticky". They can easily be joined with other fragments that were cut with the same enzyme and so have complementary sticky ends.

Stringency: Reaction conditions—notably temperature, salt, and pH—that dictate the annealing of single-stranded DNA/DNA, DNA/RNA, and RNA/RNA hybrids. At high stringency, duplexes form only between strands with perfect one-to-one complementarity; lower stringency allows annealing between strands with some degree of mismatch between bases.

Stringent plasmid: A plasmid that only replicates along with the main bacterial chromosome and is present as a single copy, or at most several copies, per cell.

Structure-functionalism: The scientific tradition that stresses the relationship between a physical structure and its function, for example, the related disciplines of anatomy and physiology.

Subcloning: The process of tranferring a cloned DNA fragment from one vector to another.

Substantial equivalence: The concept of substantial equivalence is used as a guide in the safety assessment of genetically modified foods by comparing the novel food to its unmodified counterpart which has a history of safe use. This approach allows regulatory agencies to include in their consideration, the substantial history of information related to foods which have long been safely consumed in the human diet to aid in the identification of potential safety and nutritional issues. *Definition from Health Canada Web site.*

Subunit vaccine: A vaccine composed of a purified antigenic determinant that is separated from the virulent organism.

Supercoiled plasmid: The predominant in vivo form of plasmid, in which the plasmid is coiled around histone-like proteins. Supporting proteins are stripped away during extraction from the bacterial cell, causing the plasmid molecule to supercoil around itself in vitro.

Supergene: A group of neighboring genes on a chromosome that tend to be inherited together and sometimes are functionally related.

Supernatant: The soluble liquid &action of a sample after centrifugation or precipitation of insoluble solids.

Susceptibility: When individual or species will become ill or possibly die in the face of the stressful event or circumstances.

TATA box: An adenine- and thymine-rich promoter sequence located 25-30 bp upstream of a gene, which is the binding site of RNA polymerase.

T-DNA (transfer DNA, tumor-DNA): The transforming region of DNA in the Ti plasmid of Agrobacterium tumefaciens.

Template: An RNA or single-stranded DNA molecule upon which a complementary nucleotide strand is synthesized.

Termination codon: Any of three mRNA sequences (UGA, UAG, UAA) that do not code for an amino acid and thus signal the end of protein synthesis. Also known as stop codon.

Terminator region: A DNA sequence that signals the end of transcription.

Tesla [T]: Unit of magnetic induction: 1T = 1 weber/m2 (also, 1T = 104 gauss).

Thermistor: A temperature-measuring device, that contains a resistor or semiconductor whose resistance varies with temperature.

Thermocouple: A temperature-measuring device, which contains a pair of end-joined dissimilar conductors in which an electromotive force is developed by thermoelectric effects when the joined ends and the free ends of the conductors are a different temperature.

Thiol: An SH group, or a molecule containing one. Also known as a sulfhydryl or mercapto group.

Threshold: The smallest input signal that will cause a readable change in the output signal.

Thrombin : The enzyme that converts *fibrinogen* to *fibrin* to form a blood clot. Thrombin circulates in the inactive form, prothrombin.

Thylakoid: Flattened sac of membrane in a chloroplast that contains pigment and carries out the light-gathering reactions of photosynthesis. Stacks of thylakoids form the grana of chloroplasts.

Thymidine kinase (tk): An enzyme that allows a cell to utilize an alternate metabolic pathway for incorporating thymidine into DNA. Used as a selectable marker to identify transfected eukaryotic cells.

Ti (tumor-inducing) plasmid: A giant plasmid of Agrobacterium tumefaciens that is responsible for tumor formation in infected plants. Ti plasmids are used as vectors to introduce foreign DNA into plant cells.

Toxic Substances Control Act (TSCA): See Environmental Protection Agency.

Transcapsidation: The partial of full coating of the nucleic acid of one virus with a coat protein of a differing virus. See Coat protein.

Transcription: The process of creating a complementary RNA copy of DNA.

Tanscription: The copying of genetic information from DNA to messenger RNA (mRNA).

Transduction: The transfer of DNA sequences from one bacterium to another via lysogenic infection by a bacteriophage (transducing phage).

Transfection: The uptake and expression of a foreign DNA sequence by cultured eukaryotic cells.

Transformant: In prokaryotes, a cell that has been genetically altered through the uptake of foreign DNA. In higher eukaryotes, a cultured cell that has acquired a malignant phenotype.

Transformation: (1) Transformation is the step in the genetic engineering process where a new gene (transgene) is delivered into the nucleus of a plant cell and inserts into a chromosome where it is passed on to progeny. Transformation means to genetically change a living thing. (2) In prokaryotes, the natural or induced uptake and expression of a foreign DNA sequence—typically a recombinant plasmid in experimental systems. In higher eukaryotes, the conversion of cultured cells to a malignant phenotype—typically through infection by a tumor virus or transfection with an oncogene.

Transformation efficiency: The number of bacterial cells that uptake and express plasmid DNA divided by the mass of plasmid used (in transformants/ microgram).

Transforming oncogene: A gene that upon transfection converts a previously immortalized cell to the malignant phenotype.

Transition-state intermediate: In a chemical reaction, an unstable and high-energy configuration assumed by reactants on the way to making products. Enzymes are thought to bind and stabilize the transition state, thus lowering the energy of activation needed to drive the reaction to completion.

Translation: The process of converting the genetic information of an mRNA on ribosomes into a polypeptide. Transfer RNA molecules carry the appropriate amino acids to the ribosome, where they are joined by peptide bonds.

Translocation: The movement or reciprocal exchange of large-chromosomal segments, typically between two different chromosomes.

Upload: (a) To transfer the consciousness and mental structure of a person from a biological matrix to an electronic or informational matrix (this assumes that the strong AI postulate holds). The term "Downloading" is also sometimes used, mainly to denote transferring the mind to a slower or less spacious matrix. (b) The resulting infomorph person. [The origin of the term is uncertain, but obviously based on the computer technology term 'uploading' (loading data into a mainframe computer).]

Upstream: The region extending in a 5' direction from a gene.

Vaccine: A preparation of dead or weakened pathogen, or of derived antigenic determinants, that is used to induce formation of antibodies or immunity against the pathogen.

Vaccinia: The cowpox virus used to vaccinate against smallpox and, experimentally, as a carrier of genes for antigenic determinants cloned from other disease organisms.

Variable surface glycoprotein (VSG): One of a battery of antigenic determinants expressed by a microorganism to elude immune detection.

Variation: Differences in the frequency of genes and traits among individual organisms within a population.

Vasculocyte: In medical nanorobotics, a theorized (nanorobotic) device capable of performing repairs of an injured vascular luminal surface.

Vasculography: A physical description (and mapping) of the human vascular system.

Vasculoid: Is a single, complex, multisegmented nanotechnological medical robotic system capable of duplicating all essential thermal and biochemical transport functions of the blood, including circulation of respiratory gases, glucose, hormones, cytokines, waste products, and cellular components. The vasculoid [concept] is a single, complex, multisegmented nanotechnological medical robotic system capable of duplicating all essential thermal and biochemical transport functions of the blood, including circulation of respiratory gases, glucose, hormones, cytokines, waste products, and cellular components. Vasculoid is a single, complex, multisegmented nanotechnological medical robotic system capable of duplicating all essential thermal and biochemical transport functions of the blood, including circulation of respiratory gases, glucose, hormones, cytokines, waste products, and cellular components.

Vector: An autonomously replicating DNA molecule into which foreign DNA fragments are inserted and then propagated in a host cell. Also living carriers of genetic material (such as pollen) from plant to plant, such as insects.

Vector: In cell biology, an agent (virus or plasmid) used to transmit genetic material to a cell or organism.

Vegetal pole: The end at which most of the yolk is located in an animal egg. The end opposite the animal pole.

Viral oncogene: A viral gene that contributes to malignancies in vertebrate hosts.

Viroid: A plant pathogen that consists of a naked RNA molecule of approximately 250-350 nucleotides, whose extensive base pairing results in a nearly correct double helix.

Virtual Nanomedicine: Using VR to perform surgery and other functions inside the body.

Virtual reality system: A combination of computer and interface devices (goggles, gloves, etc.) that presents a user with the illusion of being in a three dimensional world of computer-generated objects.

Virulence: The degree of ability of an organism to cause disease.

Virus: A parasite (consisting primarily of genetic material) that invades cells and takes over their molecular machinery in order to copy itself. A small replicator consisting of little but a package of DNA or RNA which, when injected into a host cell, can direct the cell's molecular machinery to make more viruses. Particle consisting of nucleic acid (RNA or DNA) enclosed in a protein coat and capable of replicating within a host cell and spreading from cell to cell. Often the cause of disease.

Viscoelasticity: A type of deformation exhibiting the mechanical characteristics of viscous flow and elastic deformation.

Viscosity: The ratio of the magnitude of an applied shear stress to the velocity gradient that it produces; in other words: a measure of a noncrystalline material's resistance to permanent deformation.

1

General Course Description of a Biotechnology Program

Biotechnology Education: An Overview

Biotechnology is the use of biological systems — living things — to make or change products. It has been used for centuries in traditional activities like baking bread and making cheese. Traditional biotechnology involved for example, developing a new wheat variety with early ripening characteristics by cross-breeding different types of wheat until the desired characteristics and only those characteristics were present in the new wheat variety. The terms 'gene technology', 'genetic engineering' and 'genetic modification' mean the same thing, and refer to one type of modern biotechnology. Since the 1940s scientists have known that DNA — deoxyribonucleic acid — in the cells of all living things is like a blueprint that is passed from one generation to another. Genes are made of DNA. They contain coded instructions for proteins, which give living things their particular characteristics like hair and eye colour. All living organisms use DNA as the genetic code. Gene technology includes a range of techniques to copy genes and modify them so that they will work in a new host either to produce or reduce a product. It also involves transfer of the reconstructed gene to a new host. Using gene technology, scientists aim to introduce, enhance or delete particular characteristics of a living thing, depending on whether they are considered desirable or undesirable. As you go further into this website, you can read more about gene technology — its history, how it's done, its possible uses, the regulatory processes in place for gene technology, the debate surrounding gene technology and how this technology might affect you.

Gene Technology

Gene technology is a term that refers to a whole range of techniques for genetic analysis that depend on the direct manipulation of DNA, the material substance of heredity. The use of gene technology in biomedicine, agriculture, food production and processing is an issue that has evoked strong public interest and concern. The debate has focused on the use of genetically modified organisms (GMOs) in food, their safety and potential impact on the environment. The public better accepts medical and industrial uses of gene technology. The majority of Australians are eager for more quality information about the technology and its applications.

Most people think of gene technology as adding genes from other life forms to plants and animals. While this is certainly part of it, the technology is broader than that, and has many more uses. Gene technology is developing at a rapid rate and it will continue to revolutionise basic biological research and development. It provides the potential to improve our health, create a safer and more secure food supply, generate greater prosperity and attain a more sustainable environment. Gene technology is already providing new ways of preventing, treating and curing human and animal diseases; it is helping farmers improve agricultural production with less impact on the environment; and in the near future, it will allow better food products to be available to consumers at reduced cost. The technology has given us vital new products like human insulin for diabetes, interferon and other drugs for treating certain cancers, and vaccines against diseases like hepatitis B. Millions of human lives are protected by gene technology every day. Genes are made from a chemical called DNA (deoxyribonucleic acid) that is common to all forms of life on Earth. Genes carry information for making all the proteins required by all organisms. These proteins determine, among other things, how the organism looks, how well its body uses food or fights infection, and sometimes even how it behaves. DNA is made up of four similar chemicals (called bases and abbreviated A,T, C, and G) that are repeated millions or billions of times throughout a genome. A genome is the entire DNA in an organism, including its genes. The human genome, for example, has 3 billion pairs of bases. The particular order of As, Ts, Cs, and Gs is extremely important. The order is called DNA sequence and variations in sequence underlie all of life's genetic diversity. DNA sequence is the database determining how the cell is to make proteins and what functions the proteins can perform. Because all organisms are related through similarities in DNA sequences, insights gained from non-human genomes often lead to new knowledge about human biology, and insights from the genes of a bacterium, worm or bee might contribute to new solutions for the wheat farmer. Gene technology consists of tools and techniques that scientists can use to study, identify or modify the genes of living organisms.

Techniques and Tools

Genetic modification

Over 20 years ago it was discovered that genes, and parts of genes, could be extracted from DNA using protein "scissors" then copied, or cloned. The gene could then be equipped with genetic "switches" to turn it up or down and inserted back into a living organism. This is the basis of genetic modification. It allows the sequences of genes to be determined. Transferring single genes between different plants and animals, turning existing genes up or down, or removing a gene from its original position and placing it in a new position in the same organism are all referred to as *genetic modification (GM)*. The terms genetic engineering (GE) or genetic manipulation also mean the same thing. Plants, animals or microbes which have a new gene inserted into them are called genetically modified organisms (GMOs) or transgenics. The modified gene belongs to the new host but its sequence is altered for example, on the basis of specific information gleaned from studies of other organisms.

DNA sequencing

DNA sequencing is the process of determining the exact order of the chemical building blocks (bases) that make up DNA. It is a laboratory procedure that involves first breaking down the DNA into short pieces, followed by separating the individual fragments using a technique called gel electrophoresis. A bar code pattern of the DNA pieces is produced which can then be read by computer. An enormous volume of information on the genetics of organisms is being generated using this technique and is providing computer experts with a great challenge in handling the data.

Human Genome Project

The Human Genome Project (HGP), which began in 1990, was a 13-year international project involving Australia. Its goal was to discover the approximate 50,000 genes that make up the human genome, make them accessible for further biological study and to determine the complete sequence of the 3 billion DNA subunits. Knowledge gained through this research will allow scientists and doctors of the future to prevent or treat most diseases at the level of genes. While it is now common to screen infants for inherited genetic diseases, further understanding of the human genome will allow doctors to determine a person's chances of developing particular diseases later in life. They will then be able to give advice or treatment to prevent, slow or cure those diseases.

Cloning Technology

A "clone" is a copy of a plant, animal, microorganism or gene derived from a single common ancestor gene, cell or organism. Identical twins arc natural clones from the one fertilised egg. A plant cutting is a clone of the original plant. Some confusion arises because the word *clone* is also applied to genes. A gene is said to be cloned when its sequence is multiplied many times in a common laboratory procedure. The cloning of a gene is a step in determining its sequence and initiating many types of experiments to understand its function and biology. The possibility of human cloning, raised when Scottish scientists at the Roslin Institute created the much-celebrated sheep "Dolly" in 1996, aroused worldwide interest and concern because of its scientific and ethical implications. Dolly was produced by injecting a nucleus from a mammary cell into an egg cell without a nucleus and raising an animal from that egg by treating it as a fertilised egg. Dolly was therefore a genetic clone of the animal from which the nucleus was taken. The important difference from identical twins is that the donor was a mature sheep and could even be made the mother of the clone. Cloning of animals and plants can be used to develop more efficient ways to produce superior breeds of animals. Cloning also enables genes to be added (such as those for human proteins) to produce animals that can generate hormones and other pharmaceuticals in milk, eggs or other products. The governments of many nations involved in gene technology have banned the cloning and genetic modification of human individuals. However, cloning technology may soon be available for human benefit to produce whole organs or special tissues from single cells for transplant to humans. It should be possible to use cells from the patient's own body and correct genetic defects before growing the tissue for transplant back into the donor. This is called therapeutic cloning.

Gene Marker Technology

DNA probes in diagnosis

Scientists can use gene technology to locate and analyse single genes in a chain of many thousands. They can make gene probes that recognise DNA sequences associated with genetic diseases. The diagnostic tests involve analysing minute tissue samples collected from adults, or taken from embryos. This information can assist parents and doctors to reduce the incidence of crippling genetic defects, and treat other disorders much better.

Forensic DNA fingerprinting

No two individuals have the same genetic makeup, a fact that is now being used by forensic scientists to identify individuals from cell samples left at a crime site. "Forensic DNA fingerprinting" involves taking a blood or tissue sample from a person and using one of a number of gene technologies to generate patterns of DNA fragments separated by size. Those patterns look something like a long bar code and are dependent on the exact sequences in the sample. A match between the genetic fingerprint from the crime sample and the suspect has been admissible evidence for a number of years.

DNA markers in breeding

For centuries plants and animals have been selectively cross-bred for desirable traits such as size, enhanced quality and pest resistance. Scientists can now use DNA information to identify desirable traits in organisms. These DNA markers are helping breeders to select superior strains of plants or types of animals for commercial production. The technique is similar to the forensic DNA fingerprint but where a particular DNA bar is known to be associated with a specific superior trait. By using these markers superior animals or plants can be identified easily and early and these individuals selected for further breeding to improve the herd or crop. CSIRO research on gene markers includes:

- markers in cattle for carcass weight, fat content, colour, meat tenderness and leanness. This information assists breeders to select superior types of animals.
- in fish farming, markers are helping breeders of super prawns select for desirable traits like size and colour, disease resistance and environmental suitability.
- in crop breeding, markers are being used to combine genes for disease resistance and quality characteristics, such as higher yield and bread making properties in wheat.

Most animals, plants and fish are still bred in the same way they have been for over 2000 years — but gene technology means that breeders can now pick the best parents with greater precision.

Transgenics

Because DNA is the same in all living things, researchers have found it possible to use genetic information derived from one organism in the modification of the genes of another. It is possible to modify the physical characteristics in precise ways. The term 'transgenics' often used to describe transferring genes between species, but in reality the process is a sharing of genetic knowledge for specific outcomes. In practice, the modified gene often has some of the gene sequences and signals from the host with some sequence information provided by the donor. Gene transfers typically involve the exact modification of one gene while leaving the other 50,000 genes unaffected. Over the last 25 years, researchers have learned how to make precise genetic changes by identifying and transferring individual genes to produce a desired characteristic. Scientists can insert new genes into an organism to add a particular trait not found in that organism. The new organism is then tested over several generations before researchers know whether the modified organism has all the intended benefits. Methods for inserting a gene into an animal or plant include:

- microinjecting into an animal egg using a fine tube
- using harmless viruses to carry genes into a host
- inserting DNA into plant cells on gold or tungsten particles
- using *Agrobacterium,* a bacterium that naturally transfers DNA into plants.

Genes are being introduced into plants to produce:

- resistance to insect attack or microbial damage, so reducing pesticide sprays and natural carcinogenic aflatoxins (mould toxins)
- tolerance to benign herbicides, so weeds can be controlled and erosion minimised through minimum tillage farming
- higher quality protein and enhanced vitamin or iron levels
- healthier vegetable oils and starches.

Genes are being inserted into animals to make them:

- produce more meat and less fat
- grow faster
- produce more meat or milk for the same amount of feed
- become more resistant to disease and parasites.

Because the techniques used to transfer genes sometimes have a low success rate, before a gene is transferred, another gene — called a "selectable marker gene" - is sometimes attached to the target gene to let researchers know whether the new gene is present or not. Scientists often use genes for antibiotic-resistance as selectable markers to show if plants have taken up a new gene. This practice allows only the genetically modified cells to grow in a special culture that contains the specific antibiotic. The antibiotics are used only in the laboratory and are of no medical significance. The antibiotic resistance genes employed are already widely distributed through the natural environment, such that their presence in the plant is of no environmental consequence. Nevertheless, because of public concerns about antibiotic-resistance genes, researchers have developed ways to remove these gene markers from plants before they are released commercially. New types of marker genes have also been developed, such as those permitting growth in the laboratory on unusual sugars.

Gene silencing

A gene that is producing undesirable characteristics in an organism can be turned down or switched off. One way this can be achieved is by inserting a second copy of the gene, or a fragment of the gene, back to front. Other ways include RNAi and "gene shears", which cause the gene message molecule to destroy itself. This cancels the effect of the undesirable gene. Gene silencing is being used to:

- protect plants from viruses
- alter the colour of ornamental flowers
- delay ripening of tomatoes and peaches so they reach the consumer in better condition
- remove allergens (proteins that cause allergies) from peanuts, soybeans and wheat
- stop undesirable browning in potatoes and raw sugar.

Gene therapy

Many human and animal diseases have now been traced to their host being born with faulty genes that produce defective proteins. These diseases include sickle cell anaemia, cystic fibrosis, the blood clotting disorder haemophilia and Down's syndrome. Modern medicine and surgery have made great strides in treating patients with genetic diseases. Scientists are now exploring the use of gene therapy to attack these diseases at their source — by correcting faulty genes. Gene therapy holds great promise for treating disease by replacing or changing a very small part of the overall genetic program of carefully selected cells, perhaps permanently, producing a cure. It aims to restore the healthy function of cells by replacing or correcting the defective gene. Gene therapy can be used to replace an abnormal gene with a normal one, to insert a missing gene, to switch off rogue genes that may cause cancers and to stop viruses multiplying within cells.

The modified cells and genes are not passed onto children. Certain technical obstacles are yet to be overcome if gene therapy is to deliver the expected benefits to treat human disease.

Among the obstacles is the lack of a delivery system, or vector, that can safely and efficiently shuttle beneficial genes into the cells of patients — and ensure they work.

Even the most advanced cell therapy techniques are still at the experimental stage in clinical trials. Further research is needed to develop safe, reliable gene therapy techniques.

Prerequisite Course

Bio-Organic Chemistry

This course provides a foundation in structural organic chemistry, acid base chemistry, chemical thermodynamics, and reaction mechanisms. Subjects include Lewis structures, atomic and hybridized orbitals, stereochemistry, inter- and intramolecular forces of attraction, neucleophilic reaction mechanisms, functional groups, and the organic chemistry of biological molecules. Please note that this course does not count toward requirements for the master's degree in biotechnology.

Bioscience for Regulatory Affairs

This course examines the fundamental underlying scientific concepts utilized in the creation and development of biomedical products. Topics to be covered include the structure and function of biomolecules such as proteins, enzymes, carbohydrates, lipids, and DNA, as well as the structure and function of cellular components such as membranes, vesicles, organelles, and the cytoskeleton. In addition, students will examine the complexities of metabolism, DNA replication, transcription, translation, signal transduction mechanisms, apoptosis, the cell cycle, and cancer.

Core Science Courses

Students must complete four core courses before they are allowed to enroll in most of the science electives. A student who has mastered some or all of the material covered in the core courses in previous academic work may request that one or more of these courses be waived. After reviewing the student's request, the program committee may approve a waiver. The student then is required to replace the waived core course(s) with science electives. Electives should be chosen in consultation with the student's adviser and should accommodate individual career goals.

Biochemistry

This course explores the roles of essential biological molecules focusing on protein chemistry, while covering lipids and carbohydrates. It provides a systematic and methodical application of general and organic chemistry principles. Students examine the structure of proteins, their function, their binding to other molecules and the methodologies for the purification and characterization of proteins. Enzymes and their kinetics and mechanisms are covered in detail. Metabolic pathways are examined from thermodynamic and regulatory perspectives. This course provides the linkage between the inanimate world of chemistry and the living world of biology.

Molecular Biology

This course provides a comprehensive overview of the key concepts in molecular biology. Topics to be covered include nucleic acid structure and function, DNA replication, transcription, translation, chromosome structure and remodeling, and regulation of gene expression in prokaryotes and eukaryotes. Extended topics to be covered include methods in recombinant DNA technology, microarrays and microRNA.

Advanced Cell Biology I

This course covers cell organization and subcellular structure. Students examine the evolution of the cell, chromosome and plasma membrane structures and behaviors, mechanics of cell division, sites of macromolecular synthesis and processing, transport across cell membranes, cell dynamics, organelle biogenesis, and cell specialization. Students also are introduced to the experimental techniques used in cell biology to study cell growth, manipulation, and evaluation.

Advanced Cell Biology II

This course is a continuation of 410.603 (Advanced Cell Biology I) and further explores cell organization and subcellular structure. Students examine cell-to-cell signaling that involves hormones and receptors, signal transduction pathways, second messenger molecules, cell adhesion, extracellular matrix, cell cycle, programmed cell death, methylation of DNA and modification of chromatic structure, and mechanisms of the cell. The involvement of abnormalities in signal transduction pathways to oncogenesis and other disease states will be stressed.

Science Elective Courses

Please note that many of the elective courses require prior completion of most core courses.

Gene Organization and Expression

Students use genetic analysis and molecular biology techniques to investigate chromosome organization, chromatin structure, functional genomics, and mechanisms of differential gene expression. Other topics include DNA methylation, silencers, enhancers, genomic imprinting, and microarray analysis.

Vaccinology

This course will cover the biological development, immunologic concepts, and methods for vaccine delivery. Specific topics include new technologies for vaccine development such as DNA vaccines, recombinant mucosal vaccines, dendritic cells for antigen delivery, novel adjuvants, and methods to increase vaccine stability. Delivery systems for vaccines, both time tested and new methodologies such as lipid-based systems, needle-free injection systems, and novel methods such as the use of genetically modified foods, will be discussed. The underlying biological role of the innate and adaptive immune systems will be explored in light of new types of vaccines and delivery systems. Finally the process of bringing vaccines to market will be covered including government oversight and licensure.

Human Molecular Genetics

In this course students learn to use the tools of modern genomics to elucidate phenotypic variation within populations. The course uses human disease (from simple Mendelian disorders to common complex disorders) to exemplify the types of studies and tools that can be used to characterize cellular pathophysiology as well as to provide genetic diagnostics and therapies. Students become facile with linkage analysis, cancer genetics, microarray analysis (oligo and DNA arrays), gene therapy, SNP studies, imprinting, dis-equilibrium mapping, and ethical dilemmas associated with the Human Genome Project.

Principles of Immunology

This course covers molecular and cellular immunology, including antigen and antibody structure and function, effector mechanisms, complement, major histocompatibility complexes, B- and T-cell receptors, antibody formation and immunity, cytotoxic responses, and regulation of the immune response. Students are also introduced to the applied aspects of immunology, which include immunoassay design and flow cytometry. Special topics include immunomodulation, immunosuppression, immunotherapy, autoimmunity, and vaccination.

Pathogenic Bacteriology

Lecture and discussion augmented by guided readings

on pathogenic bacteria, with special attention to microorganisms that cause human disease. The course is designed to impart to the student an appreciation as well as knowledge of the history, epidemiology, cultivation, morphology, serology, biochemistry, and clinical description of the major disease-producing bacteria. Discussion of therapeutic considerations and vaccination will also be included in this course.

Microbiology

This course is an overview of microorganisms important in clinical diseases and biotechnology. Students are introduced to the general concepts concerning the morphology, genetics, and reproduction of these microbial agents. Lectures focus on individual organisms with emphasis on infectious diseases, biotechnology applications, molecular and biochemical characteristics, and molecular and serological identification methods. Students will also discuss the impact biotechnology, and particularly genomics, will have on the development of antibiotics and vaccines as treatment and preventive measures.

Virology

This course covers the advanced study of viruses with regard to the basic, biochemical, molecular, epidemiological, clinical, and biotechnological aspects of animal viruses primarily and bacteriophage, plant viruses, viroids, prions, and unconventional agents secondarily. Specific areas of virology, including viral structure and assembly, viral replication, viral recombination and evolution, virus-host interactions, viral transformation, gene therapy, antiviral drugs, and vaccines, are presented. The major animal virus families are discussed individually with respect to classification, genomic structure, virion structure, virus cycle, pathogenesis, clinical features, epidemiology, immunity, and control. The viral vectors and their applications in biotechnology are discussed.

Marine Biotechnology

This course covers the application of molecular techniques to study the marine environment and obtain useful products from marine systems. Students examine recent progress in discovery of drugs and enzymes from marine microbes and macroorganisms, biodiversity, bioremediation, molecular approaches in aquaculture, the role of marine microbes in global carbon cycling, and genomics of marine organisms.

Parasitology

This course concentrates on the developmental biology, natural history, and cell and molecular biology of the major eukaryotic parasites that cause significant human morbidity and mortality in the world. The fundamental mechanisms in host-parasite relationships, diagnosis, pathogenesis, epidemiology and control strategies will be emphasized. The major focus will be *Plasmodium spp*, the causative agent of malaria (number 3 among the top 10 single disease killers in the world). Other parasites that will be studied include *Trypanosoma* (African trypanosomiasis, Chagas disease), *Schistosoma* (Schistosomiasis), *Leishmania* (Leishmaniasis) *Onchocerca* (Onchocerciasis), other blood and intestinal helminthes, tapeworms and *Pneumocystis jiroveci*. Current research in parasitology will be discussed.

Molecular Evolution and Phylogenetics

This course covers the principles of molecular evolution and phylogenetics. Topics include patterns and analyses of DNA polymorphism, genetic evolutionary trees, molecular clocks, the evolution of multigene families, gene duplication and shuffling, transposition and horizontal gene transfer, gene number and genome size, organellar and nuclear genetic markers, genetic mutation and selection, genes in populations, viral evolution, human evolution, and the theoretical background for molecular phylogenetics. Examples of each concept will be drawn from the scientific literature in epidemiology and human or animal genetics.

Agricultural Biotechnology

In this course, students are introduced to the application of recombinant DNA technology to agriculture. Studied are methods for the introduction of foreign DNA into plant and animal cells and generation of stably transformed plants and animals. Students consider specific examples of the use of transgenic plants and animals in biotechnology, which can provide protection against insects, diseases, and tolerance to specific herbicides. They also investigate how recombinant growth hormones can result in leaner meat, greater milk yield, better feed utilization, and how transgenic plants and animals can serve as bioreactors for the production of medicinals or protein pharmaceuticals. Because recombinant agricultural products are released into the environment or consumed as foods, students also need to become familiar with environmental safety issues.

Molecular Basis of Pharmacology

This course begins by reviewing receptor binding and enzyme kinetics. Various cellular receptors and their physiology are discussed as well as the pharmacological agents used to define and affect the receptor's function. Students study the pharmacology of cell surface receptors and intracellular receptors. Also considered are the drugs that affect enzymes.

Molecular and Cellular Physiology

Students in this course gain an understanding of how coordinated regulation of bodily function occurs at the molecular and cellular levels of organization. The focus is on neurons, muscles, and hormones. Specific areas covered for excitable tissue include bioelectric properties of excitable membranes, Hodgkin-Huxley ion currents, voltage-gated ion channels and their structures, synaptic transmission, excitation-contraction coupling, and contractile properties of skeletal, cardiac, and smooth muscle cells. The biotechnological connection is the pharmacological interventions to modulate functioning of excitable tissues. For endocrine physiology, there is an overview of hypothalamic, pituitary, reproductive, and other hormones. The class uses leptin and obesity as a model hormone and pathology, respectively, and examines in detail its action as a putative fat busting hormone.

System and Integrative Physiology

This course is the second half of the Physiology sequence and involves the study of organ systems and how they are regulated by the central nervous and endocrine systems. Students will learn the structure and function of the cardiovascular, respiratory, digestive, renal, and reproductive systems, as well as their pathophysiology during disease processes. We will also study metabolic physiology in the context of exercise and diet. The biotechnological connection will be how the drug interventions modulate functioning of many of these systems.

Industrial Microbiology

This course covers the principles of various processes associated with the production and recovery of different bioproducts derived from prokaryotes and eukaryotes. Topics include the classification of microorganisms, media development, instrumentation, fermentation principles, mammalian and insect cell propagation, product recovery, protein purification, and the principles of current good manufacturing practices (cGMP). Emphasis is on large-scale production methods and production of recombinant proteins for diagnostic and clinical applications.

Molecular Development

This course covers the molecular and cellular bases of development in a variety of experimental organisms with special emphasis on mammalian and human models. From the formation of germ cells, fertilization, and early embryonic development to the final formation of organs and tissues, developmental processes are considered in the context of biotechnological applications. Application possibilities include creation of transgenic animals and drug design to combat specific types of cancer. The molecular mechanisms of developmental processes and the identification of targets for therapeutic purposes are central themes.

Product Development: From Research to Manufacturing

This course provides an extensive overview of a process for development of a pharmaceutical by a biotechnology company. The course emphasizes the importance of intellectual property, the basic sciences underpinning the development of a product, and the importance of the interaction between a biotechnology company and the Food and Drug Administration (FDA). Students learn to appreciate the importance of quality control and assurance, good manufacturing practices, preclinical and clinical testing, and the lengthy regulatory processes which govern the development, manufacture, and eventual sale of biotechnological products. Hands-on solving of practical problems and guest lecturers who are experts in the field familiarize students with the intricacies of the process.

Neurobiology

This course provides a framework for understanding the molecular physiology of neuronal structure, signaling, and circuitry and how this cellular design is ultimately integrated to achieve higher cognitive functions such as perception, control of movement, learning, and memory.

The course introduces the students to various current neuroscience topics, including but not limited to membrane physiology and electrical excitability of neurons, neurotransmitters and synaptic transmission, signaling at the neuromuscular junction, cellular and higher-order aspects of perception and motor control, molecular mechanisms of neural development, and the molecular basis of learning and memory. This course places particular emphasis on the genetic and molecular bases of a wide variety of neurological and neurodegenerative diseases such as multiple sclerosis, amyotrophic lateral sclerosis, Parkinson's, and Alzheimer's.

Genes and Disease

Because of recent advances, powerful diagnostic tests now detect genetic diseases, and there is promise of gene replacement therapy. In this course students cover general genetic principles, DNA tools for genetic analysis, cytogenetics, gene mapping, the molecular basis of genetic diseases, animal models, immunogenetics, genetics of development, genetics of cancer, and treatment of genetic diseases. Molecular methods of analysis are emphasized.

Gene Therapy

Students are introduced to gene transfer, its technical evolution, and its testing through clinical studies. Gene therapy holds promise for both genetic diseases and acquired diseases such as cancer and AIDS. The health, safety, and ethical issues surrounding gene therapy are discussed, together with the review and oversight systems established to regulate this therapy. Students also consider how industry is developing these techniques, both in new start-up companies as well as in established biotechnology and pharmaceutical companies. An overview of proprietary and patent issues in gene therapy is part of the course.

Infectious Diseases

This course focuses on infectious diseases of mankind, presented in a system-by-system format. Basic principles of host defense and microbial virulence will be discussed. Practical up-to-date information on the clinical presentation, symptoms, physical findings, laboratory diagnosis, treatment, and prevention of the general array of diseases caused by bacteria and viruses will be presented. The use of antibiotics, prophylactic agents, and vaccines along with selected aspects of pathogenesis and epidemiology will be covered. More cursory coverage will be given to the fungal and parasitic agents of human disease. The student will develop a broad understanding of the many different kinds of infectious processes that our bodies are subjected to on an ongoing basis.

Emerging Infectious Diseases

This course focuses on emerging infectious diseases from many different perspectives. The maladies addressed range from diseases that have reappeared in altered genetic forms, such as the influenza virus and the West Nile virus, to the lethal hemorrhagic fever caused by the Ebola virus. Also discussed is the threat of recombinant and ancient infectious agents such as Bacillus anthracis, causative agent of anthrax, which can be used in biological warfare weapons. Opinions from noted scientists and leaders concerning emerging diseases and the prospects for battling them successfully provide scientific and social perspective.

Computers in Molecular Biology

Retrieval and analysis of electronic information are essential in today's research environment. This course explores the theory and practice of biological database searching and analysis. In particular, students are introduced to integrated systems where a variety of data sources are connected through World Wide web access. Information retrieval as well as interpretation are discussed, and many practical examples in a computer laboratory setting enable students to improve their data mining skills. Methods included in the course are searching the biomedical literature, sequence homology searching and multiple alignment, protein sequence motif analysis, and several genome analytical methods. Classes are held in a computer laboratory. Acquaintance with computers is required.

Practical Computer Concepts for Bioinformatics

This course introduces students with a background in the life sciences to the basic computing concepts of the UNIX operating system, relational databases, structured programming, object-oriented programming, and the Internet. Included is an introduction to SQL and the Perl scripting language. The course emphasizes relevance to molecular biology and bioinformatics. It is intended for students with no computer programming background, but with a solid knowledge of molecular biology.

Bioinformatics: Tools for Genome Analysis

Several large-scale DNA sequencing efforts have resulted in megabase amounts of DNA sequences being deposited in public databases. As such, the sequences are of less use than those sequences that are fully annotated. To assign annotations such as exon boundaries, repeat regions, and other biologically relevant information accurately in the feature tables of these sequences requires a significant amount of human intervention. This course instructs students on computer analytical methods for gene identification, promoter analysis, and introductory gene expression ananlysis using software methods. Additionally, students are introduced to comparative genomics and proteomic analysis methods. Students will become proficient in annotating large genomic DNA sequences. Students complete two large sequence analysis projects during the course.

Biology of HIV and AIDS

This course includes an overview of the biology and life cycle of the immunodeficiency virus, including the simian viruses (SIVs). Specific areas of HIV immunolopathogenisis are emphasized, to include HIV diagnosis, HIV-induced immune dysfunction, and therapeutic breakthroughs in the treatment of HIV-1 disease. Students become familiarized with current methods in biotechnology that have advanced our understanding of the biology of retroviruses. Special topics include international genetic variation (subtypes and clades),. HIV vaccine development, and global economic impact.

Cancer Biology

This course provides students with knowledge of the fundamental principles of the molecular and cellular biology of cancer cells. Lectures and demonstrations explain the role of growth factors, oncogenes, tumor suppressor genes, angiogenesis, and signal transduction mechanisms in tumor formation. Discussion of aspects of cancer epidemiology, prevention, and principles of drug action in cancer management is part of the course.

Protein Bioinformatics

Because the gap between the number of protein sequences and the number of protein crystal structures continues to expand, protein structural predictions are increasingly more important. This course provides a working knowledge of various computer-based tools available for predicting the structure and function of proteins. Topics include protein database searching, protein physicochemical properties, secondary structure prediction, and statistical verification. Also covered are graphic visualization of the different types of three-dimensional folds and predicting 3-D structures by homology. Computer laboratories complement material presented in lectures.

Phylogenetics and Comparative Genomics

This course will provide a practical, hands-on introduction to the study of phylogenetics and comparative genomics. Theoretical background on molecular evolution will be provided only as needed to inform the comparative analysis of genomic data. The emphasis of the course will be placed squarely on the understanding and use of a variety of computational tools designed to extract meaningful biological information from molecular sequences. Lectures will provide information on the conceptual essence of the algorithms that underlie various sequence analysis tools and the rationale behind their use. Only programs that are freely available, as either downloadable executables or as web servers, will be used in this course. Students will be encouraged to use the programs and approaches introduced in the course to address questions relevant to their own work.

Clinical and Molecular Diagnostics

This course covers basic concepts and practical applications of modern laboratory diagnostic techniques. Topics include the principles of testing methodology, quality assurance, and the application of molecular methods to the clinical and research laboratory. The test methods to be covered include nucleic acid-based methods such as hybridization, amplification, and sequencing; non-nucleic acid methods such as HPLC, GLC, and protein analysis; and technologies such as PFGE, ribotyping, RFLP, and serological testing methodologies. In addition to the test procedures, students are exposed to aspects of statistics, quality control, regulatory issues, and applications of these methods to the diagnosis and prognosis of human disease.

Biostatistics

This course introduces statistical concepts and analytical methods as applied to data encountered in biotechnology and biomedical sciences. It emphasizes the basic concepts of experimental design, quantitative analysis of data, and statistical inferences. Topics include probability theory and distributions; population parameters and their sample estimates; descriptive statistics for central tendency and dispersion; hypothesis testing and confidence intervals for means, variances, and proportions; the chi-square statistic; categorical data analysis; linear correlation and regression model; analysis of variance; and nonparametric methods.

The course provides students a foundation to evaluate information critically to support research objectives and product claims and a better understanding of statistical design of experimental trials for biological products/devices.

Clinical Trial Design and Conduct

Through a case-study approach, this course will cover the basic design issues of clinical trials. The design of a specific trial will be studied to illustrate the major design issues, such as overall design, endpoint variables, control group selection, eligibility criteria, and potential ethical issues such as the use of placebo controls.

The course also covers the planned analysis of these studies, including approaches that are central to clinical trials, such as stratified analysis, adjustment factors, and "intention-to-treat" analyses. Also to be covered are the operational aspects of clinical trial conduct (Good Clinical Practices). Group projects include a full protocol for a clinical trial of a fictitious new drug or biologic, plus the accompanying informed consent document. The semester will conclude with a mock IRB meeting to review one of the student projects.

Clinical Development of Drugs and Biologics

This course introduces students to the planning and work required to develop potential new drugs and biologics efficiently. Students gain a thorough appreciation of FDA and ICH regulations and guidelines. Because the course emphasizes the importance of planning before the execution of any of the necessary steps, lectures use a "backwards" approach, discussing the final analysis and report before developing protocols. Topics also include an overview of preclinical investigations; NDA/BLA format and content; clinical development plans; product and assay development; the IND ; and trial design, implementation, and management.

Tissue Engineering

Tissue engineering is a highly multidisciplinary field that involves cell biology, chemistry, materials science, engineering, and medicine. This course will be a survey that introduces students to the field from scientific, clinical, manufacturing, and regulatory perspectives. Roughly the first half of the course will be devoted to background material, and the second half will focus on applications. Readings will be drawn from books and journals.

Transplantation Science

This course is designed to provide exposure and insight into transplantation science. The course emphasizes the current status of organ donation, procurement programs, and histocompatibility testing. In addition, immunological consequences of organ or tissue transplantation and subsequent immuno-suppressive regimens are examined. Special attention is given to promising and innovative technologies such as tissue engineering, cardiac regeneration, stem cell technology, and xenotransplantation. Students investigate the biological and ethical issues associated with these new technologies.

Radiation Biology (Updated)

This course will review types of ionizing radiation and their differences, physical and chemical interactions of radiation with key biological molecules, effects on living matter beginning with molecular and cellular interactions and proceeding to tissue, organ, and organism levels, emphasizing the human system. Radiation's beneficial effects in cancer therapy and medicine as well as detrimental and carcinogenic effects will be discussed. Specific units will consider food irradiation, nuclear power plant accidents, radiation terrorism, everyday sources of exposure to the US population, and other practical situations involving radiation.

Methods in Proteomics

This course covers the analytical methods used to separate and characterize pharmaceutical compounds (predominantly proteins) derived through biotechnology. While emphasis is placed on the general principles and applicability of the methods, current protocols are discussed, and problem sets representing realistic developmental challenges are assigned. Topics include chromatography (HPLC, SEC, IEC), electrophoretic techniques (2-D gel electrophoresis), spectroscopic methods (UV/Vis, fluorescence, CD), analytical ultracentrifugation, microarrays, mass spectroscopy, amino acid analysis, sequencing, and methods to measure protein-protein interactions.

Epidemiology: Diseases in Populations

Epidemiology is the study of the patterns and determinants of disease in populations. It constitutes a basic science for public health and biomedical sciences and its influence can be felt daily through the presentation of data by government, academic, and industry sources. The goal of this course is to present an introduction to epidemiological methods and inferences to biotechnology professionals with little prior experience in public health. Issues in epidemiological inference and the assessment of causal relationships from epidemiological studies will be discussed, introducing the issues of bias and confounding. Throughout the course, emphasis will be on the practical use of epidemiology and lectures will be complemented by case studies and published literature. Examples will be drawn from contemporaneous issues in chronic and infectious diseases. At the conclusion of the course, students should have a greater appreciation for the role of the epidemiologic method and be able to evaluate a basic epidemiologic study, including how the study goals and research questions relate to the design, measures, and inferences.

Current Topics in Molecular and Cellular Biology

This course is a literature-driven exploration of current topics and methodologies employed in cell biology research. By closely examining both fundamental and innovative experimental approaches, as well as cutting-edge technologies, we will explore a broad range of cell biology topics, many of which most students have previously encountered, at least at some level, in the core courses. The particular subjects and technologies discussed may vary widely from semester to semester; topics may include but are not limited to expression profiling (microarray analysis),"knock-down" with RNAi, the use of transgenic and knock-out mice, proteomics and mass spectrometry, microscopy applications (epifluorescence, confocal, and/or EM), characterizing protein/ protein interactions, and detection methods for the movement of small molecules and ions. Students enrolling in this course are expected to already have some experience in critical reading and evaluation of the primary scientific literature.

Genomic Sequencing and Analysis

The completion of the human genome sequence is just the latest achievement in genome sequencing. Armed with the complete genome sequence, scientists need to identify the genes encoded within, to assign functions to the genes, and to put these into functional and metabolic pathways. This course will provide an overview of the laboratory and computational techniques beginning with genome sequencing and annotation, extending to bioinformatics analysis and comparative genomics, and including functional genomics.

Theory and Applications of Immunoassays

Antibodies are useful as molecular tools in a variety of applications in biotechnology. They can be produced quickly, inexpensively, and in very large quantities. Students will examine how antibodies can be configured as a measurement tool called an immunoassay. Students design immunoassays for their own laboratory applications such as radio- and enzyme-immunoassays, and competitive and immunometric immunoassays. They learn how to modify antibodies chemically for conjugation of reporter molecules; and they become aware that various immunoassay performance issues can affect results. Students also investigate immunoassay formats such as configuration of antibodies as biosensors and surface plasmon resonance signaling.

Microarrays and Analysis

This course will focus on the analysis and visualization of microarray data. The general aim is to introduce students to the various techniques and issues involved with analyzing multivariate expression data. Additionally, students will visualize the results in modern statistical scripting software.

Topics include detecting and attributing sources of data variability, determining differentially expressed genes with relevant statistical tests, and controlling for false positive discovery (multiple test corrections, permutations, etc.). An introduction to linear and non-linear dimension reduction methods (PCA, PLS, isometric feature mapping, etc.) and an introduction to common pattern recognition (clustering), classification, and discrimination techniques will be included. Assignments and concepts will make use of publicly available Affymetrix and cDNA microarray data sets. Examples will mostly be demonstrated in S-plus and R (publicly available) code, with some in SAS. Free demo software tools such as Minitab, Spotfire, TreeView, Expression Profiler, and web UIs will also be utilized.

Biothreat Response & Microbial Forensics (Updated)

This course introduces the methods and techniques used for biothreat detection, identification, and medical intervention. Methods to protect the health and safety of responders, the proper procedures for threat containment, decontamination, removal, and establishment of a chain of custody for evidentiary materials are all discussed. Students will learn rapid methods for biothreat detection (using various biodetection platforms) and conventional methods such as culture, PCR, DNA fingerprinting, phage typing, microarrays, and proteomics. Procedures for working in Biosafety Levels 3 and 4 environments are presented, as well as the immunological responses to biothreat exposure, and the preventive and therapeutic options available.

Science, Medicine, and Policy in Biodefense

This course provides a comprehensive introduction to the Concentration in Biodefense. Biological warfare is introduced in its historical context, followed by the properties of the most important biological threat agents, their medical consequences and treatment, diagnostics and forensics. Relevant international and domestic policy issues are explored, along with defense strategies and the nature of existing dangers to national security. Students should leave the class with a deep understanding of biological warfare and terror agents, the consequences of their potential use, and the available means of protection.

Microbial Pathogens and the Impact on National Security

This course provides an overview of the basic microbiology, pathology, detection methods, and medical treatments required in response to a biothreat attack. Methods to protect the health and safety of responders and the general public are presented, and the procedures for biothreat weaponization and delivery, collection and storage, containment and decontamination, and risk assessment are discussed. The potential biothreat agents of bacterial, viral, and agricultural significance are presented, such as, Bacillus anthracis (anthrax), Yersinia pestis (plague), variola virus (smallpox virus), hemorrhagic fever viruses (Ebola virus, dengue virus), and microbiological toxins (botulinum, superantigens). *Non-science elective for Biotechnology program.*

Applied Molecular Biology

This course will cover applied concepts and research techniques in molecular biology. It is designed for students with a good basic knowledge of molecular biology who want to study more advanced concepts and how they may be applied in biotechnology. Topics for discussion include; applied PCR techniques, vector construction and modification, affinity chromatography, two-hybrid screening, iRNA, site-directed-mutagenesis /protein engineering, microarrays, prokaryotic and eukaryotic expression systems.

Bioassay Development

This course will cover methodological approaches to bioassay development for high throughput screening. Both cell-based (cytotoxicity; cytoprotection, high content imaging, and reporter systems) and cell-free assay systems (enzyme, FRET, time resolved fluorescence, quenching assays, and immunological assays) will be included with discussion of the potential promise and pitfalls associated with each assay system. Various assay formats, visualization techniques, and current developments in assay technology will be discussed. Project management techniques will be utilized to aid in the process of assay development.

Microfluidics and Biosensors

Microfluidics (Lab-on-a-chip technology) is the miniaturization of laboratory operations for micro-scale chemistry, high-throughput drug screening, environmental sensors, biothreat detectors, forensics, clinical diagnostics and proteomics. This course will cover microfluidic implementations of bioassay development such as sample dilution, cell lysis, chromatography, solid-phase extraction, electrophoresis, nucleic acid amplification and sequencing, analyte detection, single-cell analysis, microarray design & mass spectrometry sample preparation. The materials, design, fabrication, and testing of microfluidic chips and biosensors will be discussed, with emphasis on the applications of this technology to detect microbial pathogens and cancer markers. In addition, the course will include case studies from the literature to introduce students to intellectual property issues related to microfluidic technology.

Bioperl

This course builds on the Perl concepts taught in 410.634 Practical Computer Concepts for Bioinformatics. Perl has emerged as the language of choice for the manipulation of bioinformatics data. Bioperl, a set of object-oriented modules that implements common bioinformatics tasks, has been developed to aid biologists in sequence analysis. The course will include an overview of the principal features of Bioperl and give students extensive opportunity to use Perl and the tools of Bioperl to solve problems in molecular biology sequence analysis.

Molecular Targets and Cancer

This course will investigate potential molecular targets in cancer including receptor tyrosine kinases, G-Protein coupled receptors, the TGF beta signaling pathway, cell cycle check points, kinases and phosphatases, chemokine and chemokine receptors, nuclear receptors, suppressor proteins, metastasis and angiogenesis targets, integrins, and matrix metalloproteinases. Discussion will also include topics on what defines a molecular target and the methods by which they are evaluated.

Chemical Libraries and Diversity

Chemical diversity and "pharmacological space" will be studied with an emphasis on disciplines related to drug discovery. Medicinal chemistry, natural product chemistry, focused synthetic libraries, and combinatorial chemistry will be covered. Lipinski's rules for drug-like molecules will be discussed in detail, as well as methods for chemical analysis, in silico drug design, molecular modeling, and compound storage and handling. In addition, techniques used for assessing and harnessing chemical diversity for drug discovery will be discussed.

Stem Cell Biology

This course will involve discussion and debate on current topics concerning stem cell biology and the use of stem cells in biotechnology and therapeutics. Topics will include review and discussion of developmental & cell biology, stem cell characteristics, stem cell preparation & therapeutic uses, tissue engineering, global regulatory and ethical issues, and commercialization of stem cell therapy. Current peer-reviewed literature and guest experts in the field will provide up to date information for discussion.

Comparative Microbial Genomics: From Sequence to Significance

Hundreds of Bacterial and Archaeal genomes have been completely sequenced and thousands more will follow in near future. In this course we will learn how to make sense of this vast sea of information in order to understand the diversity of microbial life on earth: transforming DNA data into knowledge about the metabolism, biological niche and lifestyle of these organisms. The use and development of bioinformatic platforms for the sensible comparison of genetic function and context are essential for work in modern microbiology. Topics covered will include methods for sequencing, gene finding, functional prediction, metabolic pathway and biological system reconstruction, phylogenomics, ontologies, and high-throughtput functional genomics. Particular attention will be paid to publically available bioinformatics resources and their proper use. Examples will be drawn from microbes of importance to human health, industry, ecology, agriculture and biodefense. Lectures and discussions are integrated with computer exercises where appropriate.

Independent Research in Biotechnology

Students in the biotechnology program have the opportunity to enroll in an independent research course. This elective course is an option after a student has completed at least eight graduate-level courses and has compiled a strong academic record. Prior to proposing a project, interested students must have identified a research topic and a mentor who is familiar with their prospective inquiry and who is willing to provide guidance and oversee the project. The research project must be independent of current work-related responsibilities as determined by the project mentor. The mentor may be a faculty member teaching in the biotechnology program, a supervisor from the student's place of work, or any expert with appropriate credentials. Students are required to submit a formal proposal for review and approval by the biotechnology program committee. The proposal must be received by the Advanced Academic Programs office no later than one month prior to the beginning of the term in which the student wants to enroll in the course. Students must meet with a member of the program committee periodically for discussion of the project's progress, and a written document must be completed and approved by the program committee and project mentor for the student to receive graduate credit. Additional guidelines can be obtained from the AAP administrative office. Download the Guidelines for Proposal. Sample IRPs. *Prerequisites : All four core courses and four elective courses.*

Laboratory Elective Courses

All the following electives are wet lab courses. Students must bring lab coat and safety glasses to all meetings of the course.

Cell Culture Techniques

This laboratory course illustrates the use of basic cell culture techniques used in bioscience research and commercial applications. Students are introduced to cell cultivation methods, including proper use of a biological safety cabinet, sterile technique, cell enumeration and media preparation, primary cultures, cultivation of cell lines, detection of contamination, cryopreservation, transfection, and an introduction to bioassays. This course is designed for students with no prior knowledge or with limited knowledge of cell culture methods.

Recombinant DNA Laboratory

This laboratory course introduces students to methods for manipulating and analyzing nucleic acids. Students gain extensive hands-on experience with plasmid purification, restriction mapping, ligations, bacterial transformations, gel electrophoresis, as well as applications of the polymerase chain reaction. This course is not recommended for students with substantial experience in these methodologies.

Recombinant Protein Expression, Production, and Analysis

This laboratory course introduces students to the construction, production, processing, and analysis of recombinant proteins from prokaryotic and eukaryotic sources. Concepts include the design, construction, and delivery of recombinant expression clones, expression of recombinant genes in host cells, protein purification, and protein analysis. Laboratory exercises use current techniques and approaches for the cloning, expression, production, purification, and analysis of recombinant proteins in bacteria and mammalian cells.

Biodefense Laboratory Methods

This laboratory course introduces students to the methods and techniques used for biothreat detection, surveillance and identification. Using bio-simulants and demonstrations, various biodetection platforms will be discussed and presented, such as point-of-detection devices and methods, laboratory based screening and identification technologies (culture, RT-PCR, ECL, immunoassays, biosensors), and high-throughput environmental surveillance methods. Statistical tools for determining diagnostic sensitivity &

specificity, and assay validity will be discussed. Laboratory practices and procedures for working in simulated Biosafety Level 3 and 4 environments will be practiced. Students will be introduced to the current bioinformatics genomic and proteomic databases used for category 3 microbial identification and characterization.

Advanced Recombinant DNA Laboratory

This course is a continuation of 410.656 Recombinant DNA Laboratory, intended for those who completed the introductory course or who have extensive laboratory experience. This second course consists of a series of integrated laboratory exercises designed to give students hands-on experience with a variety of advanced recombinant DNA techniques. Exercises include cloning and analysis of an exon from a gene; creation of mutations within structural genes through the polymerase chain reaction (or transposon mutagenesis); and examination of mutations. Students isolate total RNA and quantitate a specific message using RT-PCR, examine gene expression using microarrays, and utilize bioinformatics tools for DNA analysis.

Immunological Techniques in Biotechnology

This laboratory course introduces students to methods for analyzing the immune system. Participants gain experience with various immunologic techniques used in research and biotechnology laboratories such as immunoassays, immunofluorescence, western blot analysis, SDS-PAGE, antibody purification (protein A), and cytokine assays. Additional topics for discussion include hybridoma technology, phage antibody libraries, therapeutic monoclonal antibodies, and flow cytometry.

High Throughput Screening and Automation Laboratory

This course will utilize hands-on instruction in automated bioassay systems for high throughput screening (HTS) as an entry point to covering pertinent aspects of HTS, such as data manipulation, storage, and analysis; liquid handling robotics; microtiter plate washing, manipulation and bar coding; HTS assay detectors; and automated devices for assay setup, validation and visualization. Cost considerations, HTS amenable assay systems, and miniaturization and scale-up will also be discussed.

Enterprise and Regulatory Affairs Elective Courses

Bioethics

Students in this course analyze and discuss traditional philosophical theories regarding the nature of the moral good. They then apply these theories to critical issues and selected cases involving experiments with human subjects, organ transplantation, in vitro fertilization, the use of animals in research, the collection and publication of research data, peer review, conflicts of interest, and other topics of current concern.

Economic Aspects of Biotechnology

This course examines the economic, business, and financial forces that define and shape the biotechnology industry, with special focus on early stage and growing companies. A major focus is on decisionmaking under uncertainty, which includes how to interpret and screen various input types in the decision process and how to balance the empirical and intuitive approaches to analyzing problems arising from the nature of the biotechnology business. The course will permit the student to prepare an analytic approach to complex problems, identify and assess risks and payoffs, evaluate risks and payoffs in monetary and non-monetary terms, develop a structure for strategy and tactics, and understand how and why people make the decisions they make. Topics include defining and distinguishing the biotechnology industry, competitive forces and impact on strategy, regulation of genetic products, planning under uncertainty, system thinking and system failure, the economic environment, estimating costs and benefits, strategic components, marketing and sales, research and development, and valuing a company

Managing Biotechnology Professionals

The roles of managers and leaders within biotechnology companies undergo constant change. Biotechnology managers must engage in new and innovative problem-solving strategies; lead a diverse and global workforce; develop partnerships with other businesses, customers, and competitors; manage horizontally and across teams; and utilize technology as a competitive advantage. The student is able to address current challenges in his/her own organization and learn methods of implementing change, such as negotiation techniques and motivation. The course includes in-depth discussions of leadership skills, communication, conflict resolution, and goal integration. Students research a biotechnology organization and analyze what is working and not working within the management systems and suggest alternatives.

Marketing Aspects of Biotechnology

This course introduces students to the strategic and tactical approaches used in the marketing of biotechnological products and services. Students gain a thorough understanding of the research and planning necessary to develop a marketing plan, the relationship between the marketing and sales functions, the difference between marketing a scientific product and a scientific service, pricing strategies, distribution alternatives, communications, promotion, and the importance of perception. Knowledge of marketing terminology and techniques prove helpful to anyone in the industry.

Creating a Biotechnology Enterprise

This course provides a foundation to start or help grow a young biotechnology company from inception through early growth. Topics addressed shall include market assessment of innovative technology, patents and licensing, corporate law, preparing a business plan, raising money from angels and venture capitalists, government grants, strategic alliances, sales and marketing, real estate, human resources, and regulatory affairs. The course provides a survey and overview of the key tasks and challenges typically faced by biotech entrepreneurs, their management team, and directors. Students will prepare a business plan for a biotech start-up and present the plan to a panel of industry experts and financiers. Leaders from our local bioscience community will be guest lecturers for many of the classes.

Research Ethics

This course covers fundamental concepts in ethical decision-making in research through discussions of case studies, videos, and film. Students explore issues central to these areas, such as the appropriate use of animals in research, protections against risk in human subjects research, and topics related to integrity of the research process such as: authorship, peer review, and the ethics of the business of science.

Regulatory Processes for Domestic and Global Biotechnology Products

Regulatory affairs (RA) comprises the rules and regulations governing product development and post-approval marketing. In the U.S., the FDA establishes and oversees the applicable regulations. Biotechnology products may be classified as drugs, biologics, medical devices, or combination products. Each type is regulated by a different center within the FDA. This course provides an overview of RA and its impact on product development. Topics include RA history, regulatory agencies, how to access regulatory information, requirements for submissions and approval for drugs, biologics, and medical devices; compliance with GXP (Good Laboratory/Clinical/Manufacturing Practices), Quality Systems Regulations, ethics, and FDA inspections.

Legal Aspects of Biotechnology

In this course students gain a thorough understanding of the legal steps necessary to protect and market biotechnological inventions and of the procedures required to obtain the necessary permits and licenses from government agencies. Topics include inventorship and ownership issues in academia and industry; what can and should be patented in the United States and in other countries; how patents are granted; how to avoid losing patent rights; how to enforce and defend patents; and how to transfer rights to technology.

Bioscience Communication

Researchers must communicate effectively so their discoveries can be shared with others. In this course, students learn how to communicate their ideas to other researchers, their scientific peers, and investment communities. Students master both written and verbal communication skills; they hone their expertise at making both formal and informal oral presentations; they prepare poster presentations, and develop their own public speaking strategies. The course also presents personal strategies for improving daily communications, cross-cultural communications, and non-verbal skills. Students improve their written communication, editing, and informal writing skills. Participants also learn effective e-mail strategies for getting their message across, and learn how effective writing can improve their chances of getting grant applications approved. Class assignments include preparation of scientific papers, general science writing, oral presentations, PowerPoint presentations, and scientific posters.

Biological Processes in Regulatory Affairs

This course provides an overview of the biological processes and laboratory techniques utilized for the discovery, development and evaluation of therapeutic drugs. Students investigate drug development processes such as gene cloning, culture scale-up, downstream processing, and product purification. Emphasis is placed on the theory and application of laboratory methods used in drug development, such as recombinant DNA techniques, antibody technology, protein purification, immunoassays, high-throughput drug screening, chromatography, electrophoresis, cell receptor characterization, pharmacokinetics, drug toxicity testing and evaluation of therapeutic drugs, diagnostics, and vaccines.

International Regulatory Law

The sale of food, drugs, or medical devices generally requires prior government approval of the product and how the product will be made and sold—for example, what representations will be made on the package and the label. The legal parameters governing the approval process vary across the world, and the regulatory rules and climate in, for example, Europe, Asia, Canada, and Australia will be examined. Those procedures then will be compared and contrasted with the laws, regulations, and processes of the United States. Opportunities for collaboration and cooperation between and among countries, such as settling on guidelines for conducting clinical trials in a foreign country, will be discussed. In light of the globalization of economies, is there a possibility of harmonization of approval processes? The regulation of the importation and exportation of products and information across borders will conclude the course.

International Regulatory Affairs

The pharmaceutical/biotechnology product approval and marketing requires a good understanding of international regulatory affairs in order to successfully compete in today's global marketplace. It is important for tomorrow's leaders to understand and follow the regulatory differences to ensure optimum product development strategies, regulatory approvals and designs for exports conforming to the foreign regulatory bodies. There are various product development strategies that industry is using to shorten the product development time by conducting preclinical programs outside the United States , however, the strategy requires a careful planning and interaction with the U.S. and foreign regulatory agencies. With the increase in globalization of economy and exports, international regulations will have a bigger impact on the biotechnology business in the future. The course provides a review and analysis of the pharmaceutical/ biotechnology product approval processes within the world's major markets. The key strategies required in preclinical product development to marketing approval of the products in Europe , Japan , and the United States will be compared and discussed. Students will explore the European Union (EU) regulations and their overall importance on international markets. The course will cover the salient features of common technical and regulatory documents required for submission and approval to the leading regulatory bodies in the world, general guidance documents, International Harmonization, and General Agreement on Tariffs and Trade (GATT).

Food and Drug Law

The Food, Drug, and Cosmetic Act (FDCA) governs the regulatory approval process for bringing a drug, food, or cosmetic to market. The class will begin with discussion of the administrative procedures followed by the FDA. The course will focus on the drug and medical device approval process, but will include a brief discussion on the food approval process, what constitutes generally regarded as safe (GRAS) and the food additive petition. Application for a new drug will be described and contrasted with the abbreviated new drug application process. Over-the counter drugs are regulated in another manner, and the differences will be highlighted. Advertising and labeling of the drug also must comport with FDA guidelines. Good laboratory and manufacturing practices will be discussed, as will the inspections to ensure compliance. Students then will be exposed to the enforcement activities of the FDA, including searches, seizure actions, injunctions, criminal prosecutions, and civil penalties authorized under the FDC Act, as well as other statutes. The course will close with a discussion on the international movement of drugs, particularly in light of the varying costs of ethical pharmaceuticals.

Preparing a Successful Submission

In the United States , the FDA clears products to market through submissions by sponsors. These submissions are the pathway to market for regulated drugs, devices, biologics, and well-defined biotechnology products. Each submission type is reviewed by one or more of several different centers within the FDA. This course provides an overview of these submissions and the effect of submission type on product development. Topics include biological and biotech submissions as well as drug and medical device submissions, the review procedures and approval standards applied to each type, the guidance that applies to these submissions and any FDA inspections required for approval.

Marketing in a Regulated Environment

This course is designed to help students understand, use, and comply with the laws, regulations, and policy guidance documents governing advertising and promotion of products regulated by the Food and Drug Administration (FDA). The course provides an overview of the regulations, discussion of the FDA's advertising oversight, and FDA enforcement activities. The course content introduces students to advertising approved products. It provided insights on comparing competitor products, the need for head-to-head comparative data, dietary supplements, Rx products, OTC products, unapproved investigational products, and unapproved research products. The course also covers the different regulations from FDA, FTC, and CPSC. Labeling requirements, DoT, OSHA, and other international advertising activities are also covered.

Practicum in Bioscience Regulatory Affairs (MS BRA Only)

This integrative case-based course will focus on applying knowledge gained from previous courses in the MS Bioscience Regulatory Affairs program to actual cases from the U.S. Food and Drug Administration. For each case, students will assume the role of either a regulatory specialist, an FDA reviewer or senior-level policy-maker, or other involved stakeholders, such as a consumer group or an advocacy group. Students will be expected to research, evaluate, and present scientifically and legally justifiable positions on case studies from the perspective of their assigned roles. Students will present their perspectives to the class and be asked to debate the issues with the other students from the perspective of their assigned roles. The major responsibility of the students in this course will be to make scientifically and legally defensible recommendations and to justify them through oral and written communication.

Managerial Finance for Biotechnology I (formerly Financial Development for Biotechnology)

This course integrates the tools of financial analysis with real-world problems in the technology industries. Topics include modeling, costs and benefits, and ratio and break-even analysis. Students will read, prepare, and analyze financial statements. The difference between management financial and tax financial statements will also be covered.

Commercializing Biotechnology

This advanced course provides an integrated and practical approach to considering the principal areas of concern an entity faces when commercializing biotechnology, from creating or obtaining the technology through partnering with others to further develop and commercialize the technology, and finally selling the business or business line that incorporates that technology. The focus of this course is to highlight key junctures in a biotechnology company's evolution; help students identify key financial, management, and business issues at those junctures; and present practical alternatives for students to consider to resolve those issues. This course builds upon 410.650 Legal Aspects of Biotechnology and 410.646 Creating a Biotechnology Enterprise, but they are not prerequisites for the course.

Validation in Biotechnology

Understanding validation and applying a comprehensive validation philosophy are essential in today's biotechnology industry. First and foremost, validation allows a company to operate in compliance with the regulations and guidance set forth by FDA. Perhaps more importantly, it results in equipment, assays, and processes that are well understood and robust, less prone to failure, and more cost-effective. This course will introduce the fundamentals of validation, validation master planning, resource management, types of validation and the associated documentation, departmental roles and interaction, and the differences between commissioning and validation. Students will have an opportunity to solve real-world problems, generate actual validation documents, and develop validation program elements that balance regulatory requirements, operational needs, and business expectations.

Introduction to cGMP Compliance

There are many sources describing Current Good Manufacturing Practice. The fundamentals are published in the Federal Food, Drug and Cosmetic Act and in 21 CFR part 211. These regulations, however, only begin to describe the practices required in the pharmaceutical and biotech industries. Additional sources include guidance published by the Food and Drug Administration, EIRs, Form 483s, warning letters and podium comments from FDA speakers. Students will learn the history of the regulations and the Food and Drug Administration. Students will also learn practical solutions to the regulatory issues faced in the pharmaceutical and biotech industries today.

Technology Transfer and Commercialization

This course is an introduction to the multidisciplinary aspects involved in the process of bringing technical developments, particularly research emanating from universities and other nonprofit organizations, into commercial use. The course will provide an overview of the key policies, concepts, tools, issues, practices, and trends that are shaping the technology transfer field, with an emphasis on the life sciences sector.

Emerging Issues in Biotechnology

Biotechnology impacts the world and our social, political, and physical environment in ways many both inside and outside the industry may not fully understand or appreciate. It is critical to ensure that advances in biotechnology be accompanied by important public, political, and social considerations and discussions. This course will cover issues including domestic and global public perception of biotechnology, its benefits and risks, advances in bioagriculture and genetically modified foods, the impact of recombinant therapeutics on the pharmaceutical and healthcare industry, ways in which advances in biotechnology have and will continue to change our views of what life is, and how the political climate impacts advances in biotechnology discoveries. This highly interactive course will include thought-provoking debate and discussion with industry leaders, both proponents and opponents of biotechnology.

QA/QC for the Pharmaceutical and Biotechnology Industries

Many new quality initiatives for drugs, biotech products, and medical devices recently have been introduced. These include risk-based, science-based and systems-based assessments. Students will be presented with a comprehensive overview of the current best practices in quality assurance and quality control. Students will also be exposed to the most recent theories and expectations from the Food and Drug Administration.

Ethical, Legal, and Regulatory Aspects of the Biotechnology Enterprise

This course provides an overview of the important ethical, legal, and regulatory issues that are critical to the biotechnology industry. The course is divided into the three sections of ethics, legal issues, and regulations, each section taught by a faculty member who specializes in the area. The three sections are organized to provide basic material in a way that allows for an appreciation of how each influences the others. The section on ethics introduces students to core ethical values that guide the practice of science in the biotechnology industry. The section on legal issues focuses on key legal concepts, such as protecting inventions and intellectual property, and licensing. The section on regulation addresses the range of regulatory oversight mechanisms with which the biotech industry must comply and strategies for doing so.

Leading Change in Bioscience

As bioscience companies grow and mature, leadership needs evolve. Students will learn how to identify their company's position in the "Leadership Life Cycle" and learn how to select the right leadership capabilities based on their current organizational needs. Students will perform research to identify how the right leaders at the right time can dramatically improve organizational success. Class discussions will address the leadership needs of organizations from early stage research-based companies through fully integrated bio-pharmaceuticals. Through this course, students will be able to determine the stage a company or organization is in and will be able to recommend the critical qualities an ideal leadership team will have to ensure current and future growth and organizational success.

Drugs, Medical Devices, and Government

In bringing a food, drug, or medical device to market, Patent Office activities and Food and Drug Administration activities will come into play. The course will explore the administrative roles of each agency and those instances where an intersection or overlap of activities of the agencies occurs. For example, the patentability requirements of utility and enablement of therapeutic methods can depend on the onset of clinical trials. A major nexus occurs in the regulatory approval of a generic form of a patented ethical or "name brand" drug. The patent holder can have the term of the patent extended as compensation for the often prolonged time to obtain regulatory approval. Thus, the patent term is extended beyond the 17-year or 20-year term. Conversely, the generic manufacturer can enter the regulatory process before the patent term expires. Thus, the generic manufacturer legally can make and use the patented invention of another. That is an excused infringement. The earliest generic manufacturer that files an ANDA also may be granted a monopoly period for exclusive sale. The practical aspects associated with the interplay of those two activities will be explored as will the ethical and public policy issues raised by the two activities.

Crisis Management

This course will provide students with a fundamental understanding of crisis management, risk communications, media relations and public opinion research techniques in the context of the worldwide controversy over biotechnology. Students will be introduced to crisis management principles, strategies, tactics, and communications methods that will enable them to predict, manage, and control real-world controversies they may confront as they pursue careers in biotechnology. Course participants will work as a team to develop a biotechnology-specific crisis management plan for analysis and discussion, and will also have the opportunity to hone their communications skills by participating in practice media interviews during in-class sessions.

Non-Science Elective Courses

Proseminar Biotechnology

The Biotechnology Proseminar introduces students to issues and challenges facing leaders of public and private sector organizations, and communities seeking to achieve shared goals within the biotechnology industry. The course brings together diverse academic science and business disciplines (science, regulatory affairs, marketing, finance, legal, ethics, communications, etc). It explores how these disciplines can be used as powerful tools to create effective leadership, and productive collaborations within the industry, while improving managerial decision making. The Proseminar frames and integrates the combined MS.MBA Biotechnology content, methods, and tools of inquiry and analysis.

Biotechnology Education and Applications

Modern biotechnology techniques are currently being used in many areas such as food, agriculture, health care, forestry, the environment, minerals, oil and gas recovery and industrial processes to develop new products and processes, and to modify existing ones. Summarized below are some of the ways that modern biotechnology techniques have been applied in these areas.

Food and Agriculture: One of the most extensive applications of biotechnology has been in agriculture. Biotechnology techniques such as recombinant DNA techniques and mutagenesis have been used to develop plants with novel traits. These traits include herbicide tolerance and pest, insect and virus resistance. Biotechnology techniques have been used to produce biopesticides which are toxic to targeted plant pests. Also, experiments into genetic modification of aquatic organisms such as salmon, for such novel traits as enhanced growth, have been carried out using recombinant DNA techniques.

Health Care: To date, applications of biotechnology in health care have focused on fighting disease using the human body's own 'weapons'. Biotechnology medicines and therapies synthesize proteins, enzymes, antibodies and other substances which occur naturally in the human body, to fight infections and diseases. However, biotechnology also uses other living organisms, that is, plant and animal cells, viruses and yeasts to help produce human medicines. There are four main areas in health care in which biotechnology is currently being used: medicines, vaccines, diagnostics and gene therapy. Several biotechnology medicines have been developed and approved to treat such diseases as anaemia, growth deficiency in children, hemophilia, diabetes and others. Biotechnology companies have also developed therapeutic products for infectious agents such as HIV, hepatitis B and influenza. Biotechnology is responsible for new vaccines such as the hepatitis B vaccine which was approved in the U.S. to fight the hepatitis B virus. Biotechnology diagnostics have been used to detect a wide variety of diseases and genetic conditions. For example, gene therapy is used diagnostically to screen donated blood to protect the blood supply from HIV and hepatitis. Biotechnology diagnostics are also used in home pregnancy tests. In a home pregnancy test kit, a protein called a monoclonal antibody (MAb), binds to Human Chorionic Gonadotrophin (HCG), causing a colour change. HCG is present in a woman's urine only during pregnancy. In gene therapy, genes are used as 'drugs' to treat hereditary genetic disorders. Faulty or missing genes can be replaced to prevent the occurrence of a genetic disease. Current gene therapy is primarily experiment-based, with a few human clinical trials, such as for the treatment of cystic fibrosis, in the early stages.

Environment: Biotechnology applications in the environment focus on using living organisms primarily to treat waste and prevent pollution. Examples of these applications include bio-filtration and bio-remediation. Bio-filtration refers to the use of microorganisms to remove complex pollutants from the air emissions and waste water discharges of many manufacturing processes. Bio-remediation refers to a number of processes that use living organisms to degrade toxic waste into harmless by-products such as water, carbon dioxide and other materials. Examples of bioremediation processes include:

- *Bio-stimulation:* A technique which involves introducing nutrients to stimulate the growth of waste-eating microorganisms already present in the environment at a waste site.
- *Bio-augmentation:* A technique whereby microorganisms, not normally present in the ecosystem of a contaminated site, are added to the site to clean up pollutants. The microorganisms added may be natural or genetically modified. The organisms die when their food is used up.
- *Phytoremediation:* A technique which uses certain plants and fungi, planted at waste sites, to naturally reclaim the sites without using chemical treatment, incineration or land filling. These plants and fungi are used because they flourish by accumulating metals and other waste materials present in the soil.

Forestry: To date, only a few biotechnology-derived products and processes used in forestry have been commercialized. Most biotechnology applications in forestry are still in the research and development stage. However, some commercial applications include the development of bacterial bio-pesticides such as *Bacillus thuringiensis* (Bt), as an alternative to chemical pesticides. Tissue culture technology is also being used to produce genetically modified seedlings for forest regeneration. For example, BCRI Inc., a biotechnology company in British Columbia, is using tissue culture to produce weevil-resistant Sitka spruce. Companies are also developing bio-fertilizers such as plant growth-promoting rhizobia bacterial fertilizers. In addition, genetic engineering has been used to develop traits of agronomic interest such as pest and disease resistance, within trees.

The pulp and paper industry has used enzymes and microorganisms in their bio-bleaching and de-inking processes.

Industrial Processes: Biotechnology has been applied in a variety of industrial processes in different ways, particularly in the use of biocatalysts in manufacturing processes. Biocatalysts are substances that initiate or modify the rate of a biological process and are generally consumed in the process. Some examples of industrial processes where biotechnology has been applied include chemicals, starch/grain processing, cleaning and textile industrial processes. In the chemicals industry, biotechnology has been used to produce commodity and specialty chemicals. It is also used in the starch and grain processing industries through the use of enzymes to turn starch into glucose and fructose. Corn and other grains can be converted to sweeteners such as high-fructose corn syrup and maltose syrup, using enzymes. Biotechnology applications have been used to produce ethanol from grain. Furthermore, biotechnology is used in the textile industry for the finishing of fabrics and garments. In the pulp and paper industry, bio-pulping is used to manufacture some products.

Metals and Minerals Recovery: Biotechnology has been used in the mining industry in such processes as bio-leaching and metal bio-remediation and recovery. Bio-leaching is the use of bacteria to extract valuable metals. Metal bio-remediation and recovery refers to the use of biologically created enzymes in the degreasing process to recover metals of value.

Energy: To date, biotechnology applications in the energy industry include the production of cleaner coal and petroleum by removing sulfur and thus reducing the environmental contaminants released during combustion. Biotechnology has also been used widely to produce bio fuels such as bio-ethanol and bio-diesel. Bio fuels are alcohols, ethers, esters and other organic chemicals made from biomass such as herbaceous and woody plants, agricultural and forestry residues and industrial waste. These fuels can be used for electricity and as fuels for transportation.

Human Applications: When treating problems that arise from genetic disorder, one solution is gene therapy. A genetic disorder is a situation where some genes are missing or faulty. When this happens, genes may be expressed in unfavorable ways or not at all, and this generally leads to further complications. The idea of gene therapy is that a non-pathogenic virus or other delivery system can be used to insert a piece of DNA—a good copy of the gene—into cells of the living individual. The modified cells would divide as normal and each division would produce cells that express the desired trait. The result would be that he/she would then have the ability to express the trait that was previously absent at least partially. This form of genetic engineering could help alleviate many problems, such as diabetes, cystic fibrosis, or other genetic diseases. The potential of genetic engineering to cure medical conditions opens the question of exactly what such a condition is. Some advocates see aging and death as medical conditions and engineering problems to be solved. They see human genetic engineering as a key tool in this *(see life extension)*. The difference between cure and enhancement from their perspective is merely one of degree. Theoretically genetic engineering could be used to drastically change people's genomes which could enable people to regrow limbs, the spine, the brain. It could also be used to make people stronger, faster, smarter, or to increase the capacity of the lungs, among other things. If a gene exists in nature, it could be brought over to a human cell. Others feel that there is an important distinction between using genetic technologies to treat those who are suffering and to make those who are already healthy superior to others. There is widespread agreement that germline engineering should not currently be allowed for either therapeutic and enhancement applications, as evidenced by a recent report by the American Association for the Advancement of Science. Though theory and speculation suggest that genetic engineering could be used to make people stronger, faster, smarter, or to increase lung capacity, the AAAS report finds that there is little evidence that this currently can be done safely or without unethical human experiments.

"Transgenic organism" is now the preferred term for genetically modified organisms with extra-genome (foreign genetic) information, as opposed to "genetically engineered" or "genetically modified" organisms (which may refer to changes made within the genome such as amplification or deletion of genes). The first Genetically Engineered drug was human insulin approved by the USA's FDA in 1982. Another early application of GE was to create human growth hormone as replacement for a drug that was previously extracted from human cadavers. In 1986 the FDA approved the first genetically engineered vaccine for humans, for hepatitis B. Since these early uses of the technology in medicine the use of the GE has expanded to supply many drugs and vaccines. One of the best known applications of genetic engineering is that of the creation of genetically modified organisms (GMOs). There are potentially momentous biotechnological applications of GM, for example oral vaccines produced naturally in fruit, at very low cost. A radical ambition of some groups is human enhancement via genetics, eventually by molecular engineering. DNA sequencing is a technique which is used to identify each base in DNA. Although the costs of DNA sequencing has dropped dramatically, the NIH estimates it costs at least $10 million to sequence 3 billion base pairs - the size of the whole human genome. Although there has been a tremendous revolution in the biological sciences in the past twenty years, there is still a great deal that remains to be discovered. The completion of the sequencing of the human genome, as well as the genomes of most agriculturally and scientifically important plants and animals, has increased the possibilities of genetic research immeasurably. Expedient and inexpensive access to comprehensive genetic data has become a reality, with billions of sequenced nucleotides already online and annotated. Now that the rapid sequencing of arbitrarily large genomes has become a simple, if not trivial affair, a much greater challenge will be elucidating function of the extraordinarily complex web of interacting proteins, dubbed the proteome, that constitutes and powers all living things. Genetic engineering has become the gold standard in protein research, and major research progress has been made using a wide variety of techniques, including:

- Loss of function, such as in a knockout experiment, in which an organism is engineered to lack the activity of one or more genes. This allows the experimenter to analyze the defects caused by this mutation, and can be considerably useful in unearthing the function of a gene. It is used especially frequently in developmental biology. A knockout experiment involves the creation and manipulation of a DNA construct in vitro, which, in a simple knockout, consists of a copy of the desired gene which has been slightly altered such as to cripple its function. The construct is then taken up by embryonic stem cells, where the engineered copy of the gene replaces the organism's own gene. These stem cells are injected into blastocysts, which are implanted into surrogate mothers. Another method, useful in organisms such as Drosophila (fruit fly), is to induce mutations in a large population and then screen the progeny for the desired mutation. A similar process can be used in both plants and prokaryotes.

- Gain of function experiments, the logical counterpart of knockouts. These are sometimes performed in conjunction with knockout experiments to more finely establish the function of the desired gene. The process is much the same as that in knockout engineering, except that the construct is designed to increase the function of the gene, usually by providing extra copies of the gene or attracting more frequent transcription.
- 'Tracking' experiments, which seek to gain information about the localization and interaction of the desired protein. One way to do this is to replace the wild-type gene with a 'fusion' gene, which is a juxtaposition of the wild-type gene with a reporting element such as Green Fluorescent Protein (GFP) that will allow easy visualization of the products of the genetic modification. While this is a useful technique, the manipulation can destroy the function of the gene, creating secondary effects and possibly calling into question the results of the experiment. More sophisticated techniques are now in development that can track protein products without mitigating their function, such as the addition of small sequences which will serve as binding motifs to monoclonal antibodies.

REFERENCES

Anderlik, M. R. & Rothstein, M.A. Privacy and confidentiality of genetic information:What rules for the new science? *Annu. Rev. Genom.Hum.Genet.* 2, 401–433 (2001).

Aparicio, S. *et al.*Whole-genome shotgun assembly and analysis of the genome of *Fugu rubripes*. *Science* 297, 1301–1310 (2002).

Avery,O. T., MacLeod, C.M.& McCarty, M. Studies of the chemical nature of the substance inducing transformation of pneumococcal types. Induction of transformation by a desoxyribonucleic acid fraction isolated from *Pneumococcus* Type III. *J. Exp.Med.* 79, 137–158 (1944).

Banerjee, N. & Zhang,M.X. Functional genomics as applied to mapping transcription regulatory networks.*Curr.Opin.*

Bush, G.W., "Stem Cell Science and the Preservation of Life," *The New York Times*, August 12, 2001, p. D13.

Cohen, S. N., Chang,A.C., Boyer, H.W. & Helling,R. B.

Collins, F. & Galas,D. A new five-year plan for the US Human Genome Project. *Science* 262, 43–46 (1993).

Collins, F. S. *et al.*New goals for the US Human Genome Project: 1998–2003. *Science* 282, 682–689 (1998).

Drews, J.& Ryser, S. The role of innovation in drug development.

Druker, B. J. Imatinib alone and in combination for chronic myeloid leukemia. *Semin.Hematol.* 40, 50–8 (2003).

Fuller, B. P. *et al.* Policy forum: Ethics — privacy in genetics research. *Science* 285, 1359–1361 (1999).

Gabriel, S. B. *et al.* The structure of haplotype blocks in the human genome. *Science* 296, 2225–2229 (2002).

Gardner, M. J. *et al.* Genome sequence of the human malaria parasite *Plasmodium falciparum*. *Nature* 419, 498–511 (2002).

Gavin, A. C. *et al.* Functional organization of the yeast proteome by systematic analysis of protein complexes. *Nature* 415, 141–147 (2002).

Golub, T.R.Genomic approaches to the pathogenesis of hematologic malignancy.*Curr.Opin.Hematol.* 8, 252–261 (2001).

Guttmacher and Mark S.Guyer are at the National Human Genome Research Institute, National Institutes of Health, Bethesda, Maryland 20892, USA.

Guttmacher,A. E.& Collins, F. S. Genomic medicine — A primer.*N.Engl. J.Med.* 347, 1512–1520 (2002).

Hannon, G. J.RNA interference. *Nature* 418, 244–251 (2002).

Hilbert,D.Mathematical problems. *Bull.Am.Math. Soc.* 8, 437–479 (1902).

Hirschhorn, J.N., Lohmueller, K., Byrne, E. & Hirschhorn,K.A comprehensive review of genetic association studies. *Genet.*

Holt, R. A. *et al.* The genome sequence of the malaria mosquito *Anopheles gambiae*. *Science* 298, 129–149 (2002).

Horii, T., Nagao, Y., Tokunaga, T., and Imai, H. 20003. Serum-free culture of murine primordial germ cells and embryonic germ cells. Theriogen. 59, 1257-1264.

Horii, T., Nagao, Y., Tokunaga, T., and Imai, H. 20003. Serum-free culture of murine primordial germ cells and embryonic germ cells. Theriogen. 59, 1257-1264.

Jackson,D. A., Symons, R. H. & Berg, P. Biochemical method for inserting new genetic information into DNA of Simian Virus 40: circular SV40 DNA molecules containing lambda phage genes and the galactose operon of *Escherichia coli*. *Proc.Natl Acad. Sci.*

Kaczorowski, D.J., Patterson, E.S., Jastromb, W.E., Shamblott, M.J. 2002. Review Article. Glucose-responsive insulin-producing cells from stem cells. Diabetes Metab.

Kaczorowski, D.J., Patterson, E.S., Jastromb, W.E., Shamblott, M.J. 2002. Review Article. Glucose-responsive insulin-producing cells from stem cells. Diabetes Metab.

Kalka, C. et al. Transplantation of ex vivo expanded endothelial progenitor cells for therapeutic neovascularization. Proc. Natl. Acad. Sci. USA 2000; 97:3422-3427.

Kalkman EAJ, Bilgin YM, van Haren P, van Suylen R-J, Saxena PR, Schoemaker RG.Determinants of coronary reserve in rats subjected to coronary artery ligation or aortic banding. Cardiovasc Res 1996.

Kawano H, Do YS, Kawano Y, Starnes V, Barr M, Law RE, Hsueh WA. Angiotensin II has multiple profibrotic effects in human cardiac fibroblasts. Circulation 2000;101:1130-1137.

Kehat I, Kenyagin-Karsenti D, Snir M, Segev H, Amit M, Gepstein A, Livne E, Binah O, Itskovitz-Eldor J, Gepstein L. Human embryonic stem cells can differentiate into myocytes with structural and functional properties of cardiomyocytes. J Clin Invest. 2001; 108: 407-14.

Kellerman S, Moore JA, Zierhut W, Zimmer HG, Campbell J, Gerdes AM. Nuclear DNA content and nucleation patterns in rat cardiac myocytes from different models of cardiac hypertrophy. *J Mol Cell Cardiol* , 497-505 (1992).

Kennedy M, Firpo M, Choi K, Wall C, Robertson S, Kabrun N, Keller G. A common precursor for primitive erythropoiesis and definitive haematopoiesis.

Kerr, D.A., Lladó, J., Shamblott, M.J., Maragakis, N.J., Irani, D.N., Crawford, T.O, Krishnan, C., Dike, S., Gearhart, J.D., and Rothstein, J.D. 2003. Human embryonic germ cell derivatives facilitate motor recovery of rats with diffuse motor neuron injury. J. Neurosci. 23, 5131-5140.

Kerr, D.A., Lladó, J., Shamblott, M.J., Maragakis, N.J., Irani, D.N., Crawford, T.O, Krishnan, C., Dike, S., Gearhart, J.D., and Rothstein, J.D. 2003. Human embryonic germ cell derivatives facilitate motor recovery of rats with diffuse motor neuron injury. J. Neurosci. 23, 5131-5140.

Kimura, T., Suzuki, A., Fujita, Y., Yomogida, K., Lomeli, H., Asada, N., Ikeuchi, M., Nagy, A., Mak, T.W., and Nakano, T. 2003. Development and Disease. Conditional loss of PTEN leads to testicular teratoma and enhances embryonic germ cell production. Development 130, 1691-1700.

Kimura, T., Suzuki, A., Fujita, Y., Yomogida, K., Lomeli, H., Asada, N., Ikeuchi, M., Nagy, A., Mak, T.W., and Nakano, T. 2003. Development and Disease. Conditional loss of PTEN leads to testicular teratoma and enhances embryonic germ cell production. Development 130, 1691-1700.

Klug MG, Soonpaa MH, Koh GY, Field LJ (1996) Genetically selected cardiomyocytes from differentiating embryonic stem cells form stable intracardiac grafts. J Clin Invest 98:216-224.

Kocher A, Schuster M, Szabolcs M, Itescu S. Cardiomyocyte regeneration after neovascularization of ischemic myocardium by human bone marrowderived angioblasts. Nature Medicine 2002, in press.

Kocher AA, Schuster MD, Szabolcs MJ, Takuma S, Burkhoff D, Wang J, Homma S, Edwards NM, Itescu S. Neovascularization of ischemic myocardium by human bone-marrow-derived angioblasts prevents cardiomyocyte apoptosis, reduces remodeling and improves cardiac function. Nature Med.

L Field: Future therapy for cardiovascular disease. In Proceedings of the NHLBI Workshop Cell Transplantation: Future Therapy for Cardiovascular Disease? Columbia, MD; 1998.

Labastie M-C, Cortes F, Romeo P-H, Dulac C, Peault B. Molecular identity of hematopoietic precursor cells emerging in the human embryo. Blood 1998;92:3624-3635.

Levkau B, Koyama H, Raines EW, Clurman BE, Herren B, Orth K, Roberts JM, Ross R. Cleavage of p21cip1/waf1 and p27 kip1 mediates apoptosis in endothelial cells through activation of cdk2: role of a caspase cascade.

Li, Y., Jenkins, C.W., Nichols, M.A. and Xiong, Y. (1994) Cell cycle expression and p53 regulation of the cyclin-dependent kinase inhibitor p21. Oncogene, 9, 2261-2268 Steinman, R.A., Hoffman, B., Iro, A., Guillouf, C., Liebermann, D.A. and El-Houseini, M.E. (1994) Induction of p21 (WAF1/CIP1) during differentiation.

Liechty KW, MacKenzie TC, Shaaban AF, et al. Human mesenchymal stem cells engraft and demonstrate site-specific differentiation after in utero transplantation in sheep. Nat Med 2000, 6:1282-6.

Lin Y, Weisdorf DJ, Solovey A, Hebbel RP. Origins of circulating endothelial cells and endothelial outgrowth from blood. J Clin Invest 2000;105:71-77.

Lynch, H. T. & de la Chapelle, A. Genomic medicine: hereditary colorectal cancer.*N.Engl. J.Med.* 348, 919–932 (2003).

Lysaght, M.J., and Hazlehurst, A.L., "Private Sector Development of Stem Cell Technology and Therapeutic Cloning," *Tissue Engineering* 9(3): 555-561 (2003).

MacLellan WR and Schneider MD. Genetic dissection of cardiac growth control pathways Annu. Rev. Physiol. 2000. 62:289-320.

Mahon NG, O'Roke C, Codd MB, et al. Hospital mortality of acute myocardial infarction in the thrombolytic era. Heart 1999;81:478-82.

Maier S, Tertilt C, Chambron N, Gerauer K, Huser N, Heidecke C-

D, Pfeffer K (2001) Inhibition of natural killer cells results in acceptance of cardiac allografts in CD28-/- mice. Nature Med 7:557-562.

Makino S, Fukuda K, Miyoshi S, et al. Cardiomyocytes can be generated from marrow stromal cells in vitro. J Clin Invest 1999, 103:697-705.

Maxam, A. M. & Gilbert,W. A new method for sequencing DNA.

McCarthy PM, Rose EA, Macmanus Q, et al. Clinical experience with the Novacor ventricular assist system. J Thorac Cardiovasc Surg. 1991; 102:578-587.

McEwan PE, Gray GA, Sherry L, Webb DJ, Kenyon CJ. Differential effects of angiotensin II on cardiac cell proliferation and intramyocardial perivascular fibrosis in vivo. Circulation 1998;98:2765-2773.

Mendel, G.Versuche über Pflanzen-Hybriden.*Verhandlungen des naturforschenden Vereines, Abhandlungen, Brünn* 4, 3–47 (1866).

Milki, A.A. et al., "Incidence of monozygotic twinning with blastocyst transfer compared to cleavage-stage transfer" *Fertility and Sterility* 79(3): 503-506 (2003).

Miller, P. S. Is there a pink slip in my genes? *J.Health Care Law Policy* 3, 225–265 (2000).

National Center for Health Statistics, "Births: Final Data for 2001." National Vital Statistics Reports 51(2) (2002), available at http://www.cdc.gov/nchs/data/nvsr/ nvsr51/nvsr51_02.pdf.

National Institutes of Health Revitalization Act of 1993, Pub. L. No. 103-43, § 121(c), 107 Stat. 122 (1993), repealing 45 C.F.R. § 46.204(d).

National Institutes of Health, *Report of the Human Embryo Research Panel*, Bethesda, MD: NIH (1994).

National Research Council.*Mapping and Sequencing the Human Genome* (National Academy Press,Washington DC, 1988).

Nelissen-Vrancken H, Debets J, Snoeckx L, Daemen M, Smits J. Time-related normalization of maximal coronary flow in isolated perfused hearts of rats with myocardial infarction. Circulation 1996;93:349-355.

Nidorf SM, Siu SC, Galambos G, Weyman AE, Picard MH. Benefit of late coronary reperfusion on ventricular morphology and function after myocardial infarction. J Am Coll Cardiol 1992;20:307-313.

Nirenberg, M.W. The genetic code: II. *Sci.Am.* 208, 80–94 (1963).

Norwitz, E. R., et al., "Implantation and the survival of early pregnancy," *New England Journal of Medicine* 345(19): 1400-1408 (2001)

Ogawa M, Kizumoto M, Nishikawa S, Fujimoto T, Kodama H, Nishikawa SI. Blood 1999;93:1168-1177.

Onyango, P., Jiang, S., Uejima, H., Shamblott, M.J., Gearhart, J.D., Cui, H., and Feinberg, A.P. 2002. Monoallelic expression and methylation of imprinted genes in human and mouse embryonic germ cell lineages. Proc. Natl. Acad. Sci. USA 99, 10599-10604.

Onyango, P., Jiang, S., Uejima, H., Shamblott, M.J., Gearhart, J.D., Cui, H., and Feinberg, A.P. 2002. Monoallelic expression and methylation of imprinted genes in human and mouse embryonic germ cell lineages. Proc. Natl. Acad. Sci. USA 99, 10599-10604.

Orlic D, Kajstura J, Chimenti S, et al. Bone marrow cells regenerate infarcted myocardium. Nature 2001, 410:701-705.

Oz MC, Argenziano M, Catanese KA, et al. Bridge experience with long-term implantable left ventricular assist devices. Are they an alternative to transplantation? Circulation. 1997; 95:1844-1852.

Park, T.S., Hong, Y.H., Kwon, S.C., Lim, J.M., and Han, J.Y. 2003. Birth of germline chimeras by transfer of chicken embryonic germ (EG) cells into recipient embryos.

Park, T.S., Hong, Y.H., Kwon, S.C., Lim, J.M., and Han, J.Y. 2003. Birth of germline chimeras by transfer of chicken embryonic germ (EG) cells into recipient embryos.

Pfeffer JM, Pfeffer MA, Fletcher PJ, Braunwald E. Progressive ventricular remodeling in rat with myocardial infarction. Am J Physiol 1991; 260:H1406-14.

Pfeffer MA, Braunwald E, Moye LA, et al. Effect of captopril on mortality and morbidity in patients with left ventricular dysfunction after myocardial infarction. Results of the survival and ventricular enlargement trial. The SAVE investigators. N Engl J Med 1992;327:669-677.

Pitt B, Segal R, Martinez FA, et al. Randomised trial of losartan versus captopril in patients over 65 with heart failure (Evaluation of Losartan in the Elderly Study, ELITE). Lancet 1997;349:747-752.

Pittenger MF, Mackay AM, Beck SC, et al. Multilineage potential of adult human mesenchymal stem cells. Science 1999, 284:143-147.

Pouzet B, Ghostine S, Vilquin J-T, Garcin I, Scorsin M, Hagege AA, Duboc D, Schwartz K, Menasche P (2001) Is skeletal myoblast transplantation clinically relevant in the era of angiotensin-converting enzyme inhibitors? Circulation 104:I223-228.

Rafii S, Shapiro F, Rimarachin J, Nachman R, Ferris B, Weksler B, Moore AS, Asch AS. Isolation and characterization of human bone marrow microvascular endothelial cells: hematopoietic progenitor cell adhesion.

Ravichandran LV and Puvanakrishnan R. In vivo labeling studies on the biosynthesis and degradation of collagen in experimental myocardial myocardial infarction. Biochem Intl 1991;24:405-414.

Reich, D. E. & Lander, E. S. On the allelic spectrum of human disease. *Trends Genet.* 17, 502–510 (2001).

Rothenberg, K. *et al.* Genetic information and the workplace: Legislative approaches and policy challenges. *Science* 275, 1755–1757 (1997).

Sachidanandam, R. *et al.* A map of human genome sequence variation containing 1.42 million single nucleotide polymorphisms. *Nature* 409, 928–933 (2001).

Sanger, F.& Coulson, A. R. A rapid method for determining sequences in DNA by primed synthesis with DNA polymerase.

Sapienza, C. 2002. Commentary. Imprinted gene expression, transplantation medicine, and the "other" human embryonic stem cell. Proc. Natl. Acad. Sci. USA 99, 10243-10245.

Sapienza, C. 2002. Commentary. Imprinted gene expression, transplantation medicine, and the "other" human embryonic stem cell. Proc. Natl. Acad. Sci. USA 99, 10243-10245.

Selkoe, D. J.Alzheimer's disease: genes, proteins, and therapy. *Physiol. Rev.* 81, 741–66 (2001).

Shamblott, M.J., et al., "Derivation of pluripotent stem cells from cultured human primordial germ cells" *Proceedings of the National Academy of Sciences*. 95(23): 13726-13731 (1998).

Shi Q, Rafii S, Wu MH-D, et al. Evidence for circulating bone marrow-derived endothelial cells. Blood 1998;92:362-367.

Sidow, A. Sequence first.Ask questions later. *Cell* 111, 13–16 (2002).

Smith, L. M. *et al.* Fluorescence detection in automated DNAsequence analysis. *Nature* 321, 674–679 (1986).

Smythe GM, Grounds MD (2000) Exposure to tissue culture conditions can adversely affect myoblast behaviour in vivo in whole muscle grafts: implications for myoblast transfer therapy. Cell Transplant 9: 379-393.

Smythe G.M., Hodgetts S.I., Grounds M.D. (2001) Problems and solutions in myoblast transfer therapy. J Cell Mol Med 5:33-47.

The Chipping Forecast II. *Nature Genet.* 32, 461–552 (2002).

The Mouse Genome Sequencing Consortium. Initial sequencing and comparative analysis of the mouse genome. *Nature* 420, 520–562 (2002).

Tyson, J.J., Chen, K. & Novak, B.Network dynamics and cell physiology.*Nature Rev.Mol. Cell Biol.* 2, 908–916 (2001).

Van der Weyden, L.,Adams,D. J. & Bradley, A. Tools for targeted manipulation of the mouse genome. *Physiol.Genomics* 11, 133–164 (2002).

Wagner, K.R. Genetic diseases of muscle.*Neurol. Clin.* 20, 645–678 (2002).

Watson, J.D. & Crick, F.H. C. Molecular structure of nucleic acids: A structure for deoxyribose nucleic acid. *Nature* 171, 737 (1953).

Wilcox, A.J., et al., "Incidence of early loss of pregnancy," *New England Journal of Medicine* 319 (4): 189-194 (1988).

Xu, R.H., et al., "BMP4 initiates human embryonic stem cell differentiation to trophoblast" *Nature Biotechnology* 20(12): 1261-1264 (2002).

Zhang, M.Q.Computational prediction of eukaryotic proteincoding genes. *Nature Rev. Genet.* 3, 698–709 (2002).

2

Model Educational Programs, Courses and Sample Lessons in Biotechnology

Modd Biotechnology Education Program

The designated emphasis in Biotechnology provides a very effective multidisciplinary biotechnology concentration, which includes exposure to bioethics, business and legal aspects of biotechnology as well as a 3-6 month internship in a biotechnology company or research laboratory in another college or national laboratory. Dr. Kjelstrom, the Program Coordinator assists the students in locating internships, preparation of a strong cover letter and curriculum vitae, and the development of professional skills related to the business environment.

Affiliated Ph.D. Programs

- Agricultural and Environmental Chemistry
- Biochemistry and Molecular Biology
- Biological Systems Engineering
- Biomedical Engineering
- Biophysics
- Cell and Developmental Biology
- Chemical Engineering

- Chemistry
- Civil and Environmental Engineering
- Comparative Pathology
- Entomology
- Food Science
- Genetics
- Immunology
- Materials Science and Engineering
- Mechanical and Aeronautical Engineering
- Microbiology
- Molecular, Cellular and Integrative Physiology
- Nutritional Biology
- Pharmacology and Toxicology
- Plant Biology
- Plant Pathology
- Statistics

(A) Agricultural and Environmental Chemistry

Application Deadline

- January 15

Decrees Offered

- Ph.D.
- M.S. Plan I

Designaed Emphases

- Biotechnology
- What is a designated emphasis?

Requirements

- Three letters of recommendation
- GRE scores

Students study the chemical and biochemical aspects of foods, wine, fibers/polymers, pesticides, and environmental pollution. Research includes basic and applied problems in analytical, organic, physical, inorganic, or biological chemistry. Group research includes pesticide and toxicant chemistry, food/meat/dairy chemistry, fiber and polymer chemistry, and enzymatic reactions and fermentation chemistry.

This program is an interdisciplinary graduate group.

(B)Biochemistry and Molecular Biology

Application Deadline

- December 15

Degrees Offered

- N.S. Plan I
- M.S. Plan II
- Ph.D.

Designated Emphases

- Biophotonics
- Biotechnology
- What is a designated emphasis?

Requirements

- Three letters of recommendation
- GRE scores, including subjects test in biochemistry, biology or chemistry
- Supplemental application

Graduate work is offered in: protein chemistry, control of gene expression, gene rearrangements, chromosome structure and function, immunochemistry, molecular virology, reproductive biochemistry, structure and function of cell surfaces, protein synthesis, biochemistry of neoplasia, biochemistry of chloroplasts, lipid biosynthesis, hormonal control of metabolism, photobiology, enzymology, and membrane transport. Faculty research areas include: cell biology, molecular biology, physical biochemistry, structural biology, metabolism, and genetics in animal, plant, and microbial systems.

This program is an interdisciplinary graduate group.

For more information, visit the Biochemistry and Molecular Biology Web site.

Biological Systems Engineering

The program stresses the use of engineering to efficiently produce, distribute and process biological products, such as food, feed and fiber, while conserving natural resources, preserving environmental quality, and ensuring the health and safety of people. Programs of study include aquacultural engineering, ecological systems engineering, energy systems engineering, environmental engineering, food engineering, forest and fiber engineering, health and safety engineering, machine systems engineering, postharvest engineering, sensor and control engineering, and soil and water engineering.

This is a departmentally-based program.

For more information, visit the Biological Systems Engineering Web site.

(C) Biomedical Engineering

Application Deadline

- April 1 (domestic students); March 1 (international students)

Degrees Offered

- M.S. Plan I

- M.Engr.
- D.Engr.
- Ph.D.

Designated Emphases

- Biotechnology
- What is a designated emphasis?

Requirements

- Three letters of recommendation
- GRE scores

The program stresses the use of engineering to efficiently produce, distribute and process biological products, such as food, feed and fiber, while conserving natural resources, preserving environmental quality, and ensuring the health and safety of people. Programs of study include aquacultural engineering, ecological systems engineering, energy systems engineering, environmental engineering, food engineering, forest and fiber engineering, health and safety engineering, machine systems engineering, postharvest engineering, sensor and control engineering, and soil and water engineering.

This is a departmentally-based program.

(D) Biomedical Engineering

Application Deadline

- January 15

Degrees Offered

- M.S. Plan I
- Ph.D. Plan A

Designated Emphases

- Biophotonics

- Biotechnology
- What is a designated emphasis?

Requirements

- Three letters of recommendation
- GRE scores
- Official transcripts
- Statement of purpose

Students in this program study the effective integration of engineering with biological and medical sciences, including the modeling of biological systems, and the design of devices and procedures useful for human and veterinary medicine. A strong background in mathematics and engineering is highly recommended.

This program is an interdisciplinary graduate group.

For more information, visit the Biomedical Engineering Web site.

(E) Biophysics

Application Deadline

- January 15

Degrees Offered

- Ph.D.

Designated Emphases

- Biophotonics
- Biotechnology
- What is a designated emphasis?

Requirements

- Three letters of recommendation
- GRE scores

- Supplemental application

Biophysics offers a graduate education program that focuses on the interface of biology, physics, chemistry, engineering, and mathematics and explores the physical laws governing the properties and interaction of biomolecules and cells. The faculty has diverse research interests, which include structural biology, membrane dynamics, ion transport, electron transfer, nuclei acid, applied optics, computational biology, theory, cellular regulation, and imaging.

This program is an interdisciplinary graduate group.

(F) Cell and Developmental Biology

Application Deadline

- December 15

Degrees Offered

- M.S. Plan I
- M.S. Plan II
- Ph.D.

Designated Emphases

- Biotechnology
- Reproductive Biology
- What is a designated emphasis?

Requirements

- Three letters of recommendation
- GRE scores
- Subject test in Biology
- Supplemental application

The Cell & Developmental Biology Graduate Group includes faculty with research interests in reproductive biology and fertilization, cell motility, the cytoskeleton, intracellular transport, cytokinesis, teratogenesis, morphogenesis, the development of sensory systems, cell polarity, membrane biology and biophysics, differentiation, cellular responses to stress, metastasis, pattern formation, cell-extracellular matrix interaction, tissue engineering, signal transduction, mitochondrial DNA, DNA repair, cell adhesion, and cellular responses to growth factors. Diverse experimental systems include cell lines, the neural crest, tumor cells, yeast, vertebrates, Drosophila, C. elegans and marine invertebrates.

(G) Chemical Engineering

Application Deadline

- February 1

Degrees Offered

- M.S. Plan I
- M.S. Plan II
- Ph.D.

Designated Emphases

- Biophotonics
- Biotechnology
- What is a designated emphasis?

Requirements

- Three letters of recommendation
- GRE scores
- Minimum 600 TOEFL score for international applicants

Study in chemical engineering provides a strong background in modern chemical engineering theory and its relationship to chemical engineering processes. Areas of research include biochemical engineering, biomolecular engineering, biomedical engineering, biotechnology, catalysis, colloid and surface science, computational modeling, environmental engineering, fluid mechanics and rheology, materials processing, nanotechnology, polymers, thin films, process control, process design, separations, transport phenomena, thermodynamics, and theory and modeling.

This is a departmentally-based program.

For more information, visit the Chemical Engineering Web site.

(H) Microbiology

Application Deadline

- January 15

Degrees Offered

- M.S. Plan I
- Ph.D.

Designated Emphases

- Biotechnology
- Biology of Vector-borne Diseases
- What is a designated emphasis?

Requirements

- Three letters of recommendation
- GRE scores (subject test in biology or biochemistry is helpful but not required)
- Minimum 625 TOEFL score for international applicants.
- Strong preference is given to doctoral applicants
- Students interested in a master's degree should contact the graduate group chair prior to applying.

The microbiology group offers study in general microbiology, microbial ecology, microbial physiology, microbial genetics, molecular mechanisms of microbial regulation, molecular mechanisms of microbial pathogenesis, immunology, virology, and recombinant DNA technology.

This program is an interdisciplinary graduate group.

ADVANCEMENT FROM MOLECULAR GENETICS TO BIOTECHNOLOGY: A MODEL COURSE

This subject examines the development, impact and social context of molecular biology and genetic engineering. Topics may include: the development of a model for DNA; the development of recombinant DNA techniques; Asilomar and safety; corporate influence on molecular biology; ethical and political issues in genetic screening and genetic engineering; regulation of biotechnology and social control of research priorities; legal and moral issues in the patenting of life forms; the human genome project; the release of recombinant organisms; and biotechnology industry in Australia. On successful completion of this subject, students will have demonstrated, on the basis of written work, oral presentation and other contributions to tutorials, that they: recognise that, and can outline how, economic, political and social factors have influenced the direction of molecular biological research and determine the applications of biotechnology; can identify and illustrate issues and potential social and environmental consequences that may arise from biotechnology; can evaluate policies and means of controlling developments in molecular biology and genetic engineering; can critically analyse statements for and against the development and application of various applications of genetic engineering; and have developed their skills in finding and using arguments and information; in summarising and critically evaluating such material; and in writing and speaking.

Additional information The *Faculty of Arts*

Undergraduate Handbook 2005 is a supplement to these subject notes. See in particular material on staff consultation, student representation, assignments, assessment, Sub-Dean matters, and plagiarism. Students enrolled at Bega, Batemans Bay, Shoalhaven and Moss Vale should consult the BA (Community and Environment) Handbook for this information.

Class Topics

- designer babies
- stem cell research
- organ harvesting
- GM crops/food
- biowarfare: military and genetics
- media and genetics
- intellectual property and genetics

For each of these, we can examine ethical, social, historical, economic, political and other dimensions of the topics.

Theories

We will look at these topics through the perspectives of different social theories. Possibilities of theories include political economy, social construction of scientific knowledge, actor- network theory, participatory democracy, semiotics, feminism, consumerism, Islam and Christianity. Consult with me about other suitable theories.

Assessment

Component	*STS250*	*STS251*	*credit points*
Participation	15%	20%	1.2
Oral activity	15%	20%	1.2
Essay 1	18.75%	25%	1.5
Essay 2	26.25%	35%	2.1
Analysis	25%		2.0

Attendance

You are expected to attend at least 80% of scheduled classes. (Classes are omitted from the calculation when you provide certificates showing that absence was on medical or compassionate grounds.) If you are present for only part of a class, that counts as fractional attendance. For those with less than 80% attendance, a corresponding percentage will be subtracted from the overall mark. For example, with 72% attendance, 80% - 72% = 8% will be subtracted from your final mark.

Group Work

You are encouraged to work in groups on any of the assessment items. However, no more than 30% of your total assessment can be for collective submissions. For example, you can submit essay 1 as a group (up to three members). You might work in a team to prepare for essay 2 but submit separate essays. Discuss options with me.

Science, Technology & Society, University of Wollongong

Participation

On designated weeks, beginning with week 2, you should bring to class one or more relevant items from the media, scholarly journals, or other sources — articles, video or audio clips, emails, etc. — and be prepared to introduce them to others in the class in a few sentences.

Participation will be assessed using these criteria (roughly equally)

- relevance of the items to themes covered in the subject
- diversity and originality of items
- insight offered in introductions of items

Oral Activity

You can either make a formal presentation to the class or design an activity to help others understand the topic in its social context. To deal with the "social context," you should explicitly use a theory, as outlined above.

For the activity, you can use participatory techniques such as debates, quizzes, role plays and small group exercises.

Consider using overheads, handouts, tapes and videos. Try to relate your topic to the experiences of class members. You should also include some method to evaluate how well everyone has understood what you are trying to get across. Time allocated for each oral activity will be 20 minutes per person, e.g. 60 minutes for a three-person group.

Oral activities will be assessed using these criteria (roughly equally)

- knowledge of topic and theory
- speaking performance (presentation only)
- quantity and quality of audience involvement (activity only)
- aids (handouts, overheads, posters, videos, etc.)
- methods of evaluating how well class members have understood your message.

Essay 1

Pick one of the class topics, or another topic with my agreement. Write an opinion piece in the style of a feature article in the "Opinion" page in *The Australian* or the "Comment" page in the *Sydney Morning Herald*. You should use a theory to underpin or help motivate your comment. However, you don't need to refer to the theory explicitly. Instead, use footnotes to cite references on the topic (at least 2) and on the theory (at least 2). (When submitting articles to newspapers, it can be useful to include references to establish your credibility. The newspaper won't publish them, though.)

Length: no more than 1000 words.

Due date: 9.30am, Wednesday 6 April.

Where: In class. If you're unable to attend, put the essay under the door to my office (19.2016) or in my mail box in room 19.1048. Alternatively, mail the essay to me by express post no later than 9.30 on the due date.

Resubmission Anyone whose essay does not receive a mark of 50 or more will have the option of resubmitting it, to bring the mark up to 50.

Essay 1 will be assessed using these criteria (roughly equally)

- understanding of the topic
- understanding of the theory and its application to the topic
- persuasiveness of your argument
- quality and appropriate style of writing

Essay 2

Pick one of the class topics, or another topic with my agreement. Write an information leaflet that might be used by an organisation that is advocating a position, for example a biotechnology corporation, an anti-GM group, a church or a government department (in Australia or elsewhere). You should use a theory to underpin or help motivate your comment. However, you don't need to refer to the theory explicitly. Instead, use footnotes to cite references on the topic (at least 3) and on the theory (at least 3).

Length: 1500 words.

Due date: 9.30am, Wednesday 25 May.

Where: In class. If you're unable to attend, put the essay under the door to my office (19.2016) or in my mail box in room 19.1048. Alternatively, mail the essay to me by express post no later than 9.30 on the due date.

Resubmission Anyone whose essay does not receive a mark of 50 or more will have the option of resubmitting it, to bring the mark up to 50.

Essay 2 will be assessed using these criteria (roughly equally)

- understanding of the topic
- understanding of the theory and its application to the topic
- persuasiveness of your argument
- quality and style of writing

Analysis (STS250 only)

Choose a substantial scholarly article, a major film or some other such cultural product published since 2004 dealing with genetics or biotechnology. Analyse it using two theories. Submit or make available a copy of the product along with your analysis.

Length: 1500 words.

Due date: noon Wednesday 15 June, either under my office door (room 19.2016), in my mail box (room 19.1048). Alternatively, mail the report to me by express post no later than noon 15 June.

Analyses will be assessed using these criteria (roughly equally)

- understanding of the cultural product
- understanding of the theories
- application of the theories in analysing the product
- argument (starting from clearly articulated premises; mobilising evidence and logic towards a conclusion; recognising assumptions made and limitations)
- quality of expression (organisation of material, clarity, grammar, spelling, etc.)

Alternatively, you may do an oral presentation no later than week 12, with the same assessment criteria as for the oral activity.

SAMPLE BIOTECHNOLOGY LESSON FOR TEACHERS

In setting up the stations around the room, you may choose to present a "contrasting" food at some of the stations. This food would be one that was produced without any biotechnology application. For example, at station one, you could present matzo along- side the leavened bread as a comparison of the changes which can be attributed to the biotechnology application.

Procedure

Teacher sets up stations around the classroom prior to the beginning of class. Each station will depict a specific time period through a food representing the biotechnology applications of that time period. The five time periods to be represented are:

- B.C.
- 1 A.D.–1900 A.D.
- 1900 A.D.–1970 A.D.
- 1970 A.D.–1996 A.D.
- Future

The individual time period stations should be set up as follows:

- food sample of the time period, enough for each student to sample the food. For example, B.C. time period would have a plate of leavened bread slices.
- any appropriate cup, plate, or utensil needed to sample the food.
- informational 3 x 5 card: Each time period station has a 3 x 5 information card which discusses the link between food and biotechnology for that time period.
- Student Questions on a 3 x 5 card: Each time period station has a 3 x 5 card which has several questions for students to answer.

When the students enter the classroom, they will be:

- given a brief introduction to the activity,
- divided into five groups (group size dependent on class size),
- provided with the student answer sheet to fill in as they travel through the stations.

Each group will begin at a different station. The group will sample the food, read about the associated food/ biotechnology link and answer the student questions. Students should have approximately five to ten minutes at each station. Teacher should announce when it is time to switch stations. Groups should move in a sequential direction so that all groups get to all five stations within the class period.

Class Discussion

At the next class period, the teacher asks each group to

present to the class the answers to the questions from a specific time period.

1. Would you consider the ancient Egyptians to be biotechnologists? Why? Why not?

This question is designed to provoke students to think about what biotechnology is and what it entails. At this point in the unit, students cannot comprehend the full scope of biotechnology, but they can make judgements about what makes a person a biotechnologist. In fact, the Egyptians were biotechnologists because they manipulated a biological system to make a product. In general, biotechnology can be defined as the use of living organisms to make a product or run a process.

2. How do you think yeast causes bread to rise?

A wide variety of answers are possible here, but the actual process should be explained at some point. Yeast, a fungus of the genus *Saccharomyces,* has been used as a leavening agent for more than six thousand years. Yeast metabolizes sugar in bread dough under anaerobic conditions and converts the sugar to carbon dioxide. As yeast releases carbon dioxide, the gas expands the gluten network in the dough, displacing volume and making the dough rise.

3. What do you think the Latin root "bio" means? Define the word "technology".

"Bio" means life. "Technology" means the application of knowledge, process, invention or method for practical ends.

SAMPLE BIOTECHNOLOGY LESSON FOR STUDENTS

Station #2

Summary

This activity serves as an introduction to the entire unit on biotechnology and food. Here, students will gain an appreciation for the age and diverse scope of biotechnology by observing applications to food items throughout a long history of humankind's utilization of living systems in food preparation and production. Stations will be set up around the classroom. At each station will be a food of a specific time period which has an associated biotechnology application. A brief description of the food/technology association and related questions for students will also be at each station.

Objectives

- Students will gain an appreciation of the ubiquitousness of biotechnology applications in food production and processing.
- Students will gain some perspective as to when various techniques of biotechnology were introduced.
- Students will brainstorm as to how living systems/ organisms function to alter a food product.

Materials

- food item for each time period. The following are the foods suggested in this activity. However, there are many other appropriate examples. Teachers may want to customize this part of the activity according to appropriate foods that are readily available and not too expensive:
  - — B.C. time period: leavened bread
  - — 1 A.D. –1900 A.D. : peas
  - — 900–1970: corn (hybrids are the focus here)
  - — 1970–1996: milk
  - — Future: tomatoes, peanuts, potato chips, popcorn (any or all of these may be represented for this time period)

— paper plates, cups, bowls, and utensils as needed for each station.

— informational 3 x 5 card for each time period

— Challenge Questions 3 x 5 card for each time period.

— Student Answer Sheet

— Teacher Additional Background Information Sheet

DINING OF DNA SAMPLE LESSON

Station #2

1. Some traits are dominant over others. However, simply because a trait is dominant does not necessarily mean it is desirable. If you were a plant breeder and wanted your plants to express (have the phenotype of) a recessive trait, how would you conduct your breeding experiments?

Identify plants that express the recessive trait. They have two alleles coding for the recessive trait. By breeding these two plants, expression of the recessive trait is almost guaranteed.

2. For the following foods, list one characteristic that may have been "bred" for in that food:

oranges

grapes

turkeys

oranges: seedlessness, size, sweetness

grapes: seedlessness, size, sweetness, color

turkeys: large breast section, overall more meat, size

Station #3

1. If you could mix any two plants to form a hybrid,

what two plants would you mix? Why these two? What name would you give the hybrid?

Any thoughtful answer would be appropriate here. A good "supermarket" example is the tangelo, a mix between tangerines and grapefruit.

2. *What food(s) have you eaten that may be considered to be a hybrid(s)?*

Again, the tangelo example is appropriate here. By definition, however, many high-quality foods are hybrids because they have been bred to contain the best traits from their two parents. Therefore, some students may answer this question with foods like hybrid corns, peas, barley, wheat, etc.

MODEL SHEET FOR FOR STUDENT CHALLENGE QUESTIONS

Station 1: Time Period = B.C.

1.

2.

3a.

3b.

Station 2: Time Period = 1 A.D.—1900 A.D

1.

2. Oranges:

grapes:

turkeys:

Station 3: Time Period = 1900-1970

1. What two plants and why these two?

Name:

2.

Station 4: Time Period = 1970-1996

1.

2.

Station 4: Time Period = FUTURE

1.

2.

3.

4.

5.

Final Question:

All of the foods at the stations were produced using some types of biotechnology. In your own words, define biotechnology.

Station 4: Time Period 1970–1996

Technology: Bovine Somatotropin (BST)

Somatotropin is a mammalian hormone produced in the anterior pituitary gland beneath the brain. This hormone regulates growth and affects the metabolism of all classes of nutrients. Insufficient somatotropin production in humans leads to dwarfism, but medically advanced countries do not have a problem with dwarfism because babies diagnosed for insufficient somatotropin production may be treated with human somatotropin injections. Somatotropin is not orally active, so it is administered directly into the circulatory system. Many farmers inject their dairy cows with somatotropin, called bovine somatotropin. An abundance of the hormone in dairy cattle increases milk yield by 10%-20%, increases productive efficiency (measured by kilogram weight gain per kilogram feed) by 15%-35%, and reduces fatty tissue by nearly 80%.

The Food and Drug Administration, the National Institutes of Health, the United States Congress Office of Technology Assessment, and the American Academy of Pediatrics have determined that bovine somatotropin (BST) use is safe for three reasons:

- BST is a protein, and all plant and animal protein is degraded into single amino acids in the stomach;
- nonprimate somatotropin does not affect humans; and
- the cooking process (pasteurization) denatures somatotropin and renders the hormone biologically inactive.

Today, most of the BST administered to dairy cows is manufactured in the laboratory via recombinant DNA technology. This allows for an adequate supply of the hormone for all of the dairy farmers who wish to utilize it. The students have not been introduced to recombinant DNA technology at this time, however, you may choose to include the fact that BST is a product of a more involved technology.

Station 5: Time Period FUTURE

More and more researchers are looking into new ways of applying techniques of modern molecular biology to practices in food production, food processing and agriculture. It is projected that by the year 2030, the world's population will triple, but the number of cultivated areas will not even double. This strain on agricultural resources will undoubtedly result in greater world hunger (lack of food security). Modifying plants to withstand environmental hardships such as heat and drought has the potential for more plentiful, higher quality crop yields. The development of pest-resistant crops may help reduce overall chemical stress on the environment by reducing the overall applications of chemical pesticides.

Creating more nutrient dense forms of staple foods in third world countries has the potential to lessen disease states and enhance overall wellness of the world's poor.

Yet, the questions remain, are we looking at the whole picture? Will the practices of "genetic engineering" in agriculture negatively impact ecosystems by limiting biodiversity? Could plants engineered to contain virus particles in order to ward off pests pose a risk by having the potential to create a new virus which could then go on to negatively impact that or another crop? These questions are, as of yet, unanswered, but thought provoking. The journey in which you will be investigating food biotechnology promises to be one laden with provocative questions and opportunities for creative problem-solving. Keep in mind that food biotechnology is a subject area that can be approached from many tacks such as history, economics, environmental science, molecular biology, ethics, governmental/legislative issues, food safety, and more. Many opportunities exist for scintillating discussion regarding these issues. Enjoy!

Defining Biotechnology

It is strongly recommended that collection agencies provide survey respondents with both the single definition of biotechnology and the list-based definition. It is further recommended that statistical agencies provide an "Other (please specify)" category when using the list-based definition categories as question items. This will allow respondents to report biotechnology techniques that fit the single but not the list-based definition and will thus assist in updating the list-based definition.

The provisional single definition of biotechnology is deliberately broad. It covers all modern biotechnology but also many traditional or borderline activities. For this reason, the single definition should always be accompanied by the list-based definition which operationalises the definition for measurement purposes. The single definition is:

> *The application of science and technology to living organisms, as well as parts, products and models thereof, to alter living or non-living materials for the production of knowledge, goods and services.*

The following list of biotechnology techniques functions as an interpretative guideline to the single definition. The list is indicative rather than exhaustive and is expected to change over time as data collection and biotechnology activities evolve.

The list based definition of biotechnology techniques:

DNA/RNA: Genomics, pharmacogenomics, gene probes, genetic engineering, DNA/RNA sequencing/synthesis/amplification, gene expression profiling, and use of antisense technology.

Proteins and Other Molecules

Sequencing/synthesis/engineering of proteins and peptides (including large molecule hormones); improved delivery methods for large molecule drugs; proteomics, protein isolation and purification, signaling, identification of cell receptors.

Cell and tissue culture and engineering: Cell/tissue culture, tissue engineering (including tissue scaffolds and biomedical engineering), cellular fusion, vaccine/immune stimulants, embryo manipulation.

Process biotechnology techniques: Fermentation using bioreactors, bioprocessing, bioleaching, biopulping, biobleaching, biodesulphurisation, bioremediation, biofiltration and phytoremediation.

Gene and RNA vectors: Gene therapy, viral vectors.

Bioinformatics: Construction of databases on genomes, protein sequences; modelling complex biological processes, including systems biology.

Nanobiotechnology: Applies the tools and processes of nano/microfabrication to build devices for studying biosystems and applications in drug delivery, diagnostics etc.

Gene Technology and Biotechnology

Biotechnology is the use of plants, animals and microorganisms to create products or processes. Traditional applications include animal breeding, brewing beer with yeast, and cheese making with bacteria. Recent developments include the use of enzymes or bacteria in a wide range of applications, including waste management, industrial production, food production and remediation of contaminated land. Modern biotechnology also includes the use of *gene technology,* which allows us to move genetic material from one species to another. Biotechnology offers substantial economic, human health, agricultural, and environmental benefits, and has *applications* in a range of sectors of industry including mining, manufacturing, agriculture, biodiversity conservation and environmental management. Applications of modern biological techniques also pose potential risks to the environment. There is a need to ensure that any environmental risks associated with the technology are identified, assessed and managed through rigorous *regulatory controls.*

Environmental Applications of Biotechnology

There is huge potential for using biotechnology to reduce human impact on our environment. *Environmental uses* are still being developed, but include bioremediation (treating contaminated soils), bioprocessing (cleaner production, waste management, biomonitoring and immunocontraceptives for feral pests.

Gene technology is a category of biotechnology that involves altering the genetic material of living things to produce new or modified substances, or perform new or changed functions. While there are many benefits associated with biotechnology, including gene technology, there are also some risks that need to be assessed.

Biological control is the control of pests by natural predators, parasites, disease-causing bacteria or viruses. For example, research is being carried out to find a way of reducing the rabbit population by inserting a new gene into the myxoma virus.

Medical Applications of Biochips

Implanted biochips could be the tool of Big Brother, but they are more likely to become the treatment of choice for the physician of the 21st century. Paranoid ranting is a staple on the internet, and the topic of biochips is a particular favorite. Biochips - any microprocessor chips that can be used in biology - mean many things to many people. The web's preachers focus on a simple idea: an implanted chip that identifies each person uniquely, and can be used to track their location. This wonderful idea - capable of locating lost children, downed soldiers, and wandering Alzheimers patients - sounds alarming even to many mild-mannered civil libertarians. To Living Truth Ministries of Austin, TX, it is no less than "a tremendous bonanza for the coming Antichrist!" Second Advent Ministries of San Bernardino, Calif., warns us that the US social security number will soon be replaced (at the orders of the "New World Order United Nations-controlled world government", no less) by a chip-encoded "18-digit numbering system [which] will consist of THREE separate sets of SIX DIGITS each: 6-6-6! Coincidental? No. Not if you believe Bible prophecy." Reality is somewhat less alarming. A simple ID chip is already walking around in tens of thousands of individuals, but all of them are pets. Companies such as AVID (Norco, Calif.), Electronic ID, Inc. (Cleburne, Tx.), and Electronic Identification Devices, Ltd. (Santa Barbara, Calif.) sell both the chips and the detectors. The chips are the size of an uncooked grain of rice, small enough to be injected under the skin using a syringe needle. They respond to a signal from the detector, held just a few feet away, by transmitting out an identification number. This number is then compared to database listings of registered pets. Daniel Man, a plastic surgeon in private practice in Boca Raton, Fla., holds a patent on a more sinister device: a chip that would enable lost humans to be tracked by satellite. He has a large-scale model, but for miniaturization and regulatory approval he will need a substantial amount of cash. "There are some companies that are interested, but nothing is official," he says. "We're not giving up on it yet."

The civil liberties debate over biochips has obscured their more ethically benign and medically useful applications. Medical researchers have been working to integrate chips and people for many years, often plucking devices from well-known electronic appliances. Jeffrey Hausdorff of the Beth Israel Deaconess Medical Center (Boston, Mass.) has used the type of pressure-sensitive resistors found in the buttons of a microwave oven as stride timers. He places one sensor in the heel of a shoe, and one in the toe, and adds a computer to the ankle to calculate the duration of each stride. "Young, healthy subjects can regulate the duration of each step very accurately," he says. But elderly patients prone to frequent falls have extremely variable stride times, a flag that could indicate the need for more strengthening exercises or a change in medication. Hausdorff is also using the system to determine the success of a treatment for congestive heart failure. By monitoring the numbers of strides that a person takes, he can directly measure the patient's activity level, bypassing the often flawed estimate made by the patient. Hausdorff's chips are external, but a chip proposed by Sensors for Medicine and Science, Inc. (S4MS; Germantown, Md.) will be injected under the skin. The chip will allow diabetics to easily monitor the level of the sugar glucose in their blood. Diabetics currently use a skin prick and a hand-held blood test, and then medicate themselves with insulin depending on the result. The system is simple and works well, but the need to draw blood means that most diabetics don't test themselves as often as they should. Although they may get away with this in the short term, in later life those who monitored infrequently suffer from blindness, loss of circulation, and other complications. The solution is more frequent testing, using a less invasive method. Various companies are trying to extract the required body fluid, using either an electric current (Cygnus, Inc., Redwood City, Calif.) or a tiny pore induced by a laser (SpectRx, Norcross, Ga.) or a small needle (Integ, St Paul, Minn.). "All of these approaches ameliorate the drawing of blood," says George Daniloff, Chief Scientific Officer of S4MS, "but they still require manipulation of skin, and the skin may be perturbed." The S4MS chip, however, will sit underneath the skin, sense the glucose level, and send the result back out by radio-frequency communication. A light-emitting diode (LED) starts off the detection process. The light that it produces hits a fluorescent chemical: one that absorbs incoming light and re-emits it at a longer wavelength. The longer wavelength of light is then detected, and the result is sent to a control panel outside the body. Glucose is detected because the sugar reduces the amount of light that the fluorescent chemical re-emits. The more glucose there is, the less light that is detected. S4MS is still developing the perfect fluorescent chemical, but the key design innovation of the S4MS chip has been fully worked out. The idea is simple: the LED is sitting in a sea of the fluorescent molecules. In most detectors the light source is far away from the fluorescent molecules, and the inefficiencies that come with that mean more power and larger devices. The prototype S4MS chip uses a 22μW LED, almost forty times less powerful than the tiny power-on buttons on a computer keyboard. The low power requirements mean that energy can be supplied from the outside, by a process called induction. The fluorescent detection itself does not consume any chemicals or proteins, so the device is self sustaining. Daniloff has a working model of an oxygen sensor that uses the same layout. With it's current circuitry it is about the size of a large shirt-button, but the final silicon wafer will be less than a millimeter square. "The scale transition is not a challenge," says Daniloff. "The challenge is to illustrate the principle." The oxygen sensor will be useful not only to monitor breathing in intensive care units, but also to check that packages of food, or containers of semiconductors stored under nitrogen gas, remain airtight. Tom Ferrell of the Oak Ridge National Laboratory in Oak Ridge, Tenn., is also developing an oxygen-sensing chip, but his version sends light pulses out into the body. The light is absorbed to varying extents, depending on how much oxygen is being carried in the blood, and Ferrell's chip detects the light that is left. The rushes of blood pumped by the heart are also detected, so the same chip is a pulse monitor. A number of companies already make large-scale versions of such detectors, and Ferrell says he has "collaborated with almost every company in this area," although he will not name names. Ferrell estimates that the oxygen chip is two years away, but he has already reduced the dimensions of his temperature-sensing chip to 3mm per side. The transition of certain semiconductors to their conducting state is inherently sensitive to temperature, so designing the sensor was simple enough. With some miniature radio-frequency transmitters, and foam-rubber earplugs to hold the chip in place, the device is complete. Ferrell sees applications for anyone from sick children, to chemotherapy patients who can be plagued by sudden rises in body temperature in response to their anti-cancer drugs.

Brain Surgery with an On-off Switch

Sensing and measuring is one thing, but can we switch the body on and off? Heart pacemakers use the crude approach: large jolts of electricity to synchronize the pumping of the heart. The electric pulses of the Activa implant, made by Medtronic, Inc. (Minneapolis, Minn.), are directed not at the heart but the brain. They turn off brain signals that cause the uncontrolled movements, or tremors, associated with diseases such as Parkinson's. Drug therapy for Parkinson's disease aims to replace the brain messenger dopamine, a product of the brain cells that are dying. But eventually the drug's effects wear off, and the erratic movements come charging back. "I've had patients tell me that they scare their grandkids because they are flaying around like a fish," says Gary Heit, a neurosurgeon at Stanford University Medical Center. "Imagine your whole body twisted like a charleyhorse." The next option is to remove a portion of the thalamus, a brain structure that triggers the tremors. The implant, cleared for use in the United States in August 1997, is a new alternative that uses high-frequency electrical pulses to reversibly shut off the thalamus. "It has the same benefits as thalamotomy [the brain operation] without the risks," says Heit. The implantation surgery is far less traumatic than thalamotomy, and if there are any post-operative problems the stimulator can simply be turned off.

Adding Back the Senses

The Activa implant interferes with aberrant brain

functioning. The most ambitious bioengineers are trying to add back brain functions, restoring sight and sound where there was darkness and silence. The success story in this field is the cochlear implant. Most hearing aids are glorified amplifiers, but the cochlear implant is for patients who have lost the hair cells that detect sound waves. For these individuals, no amount of amplification is enough. The cochlear implant delivers electrical pulses directly to the nerve cells in the cochlea, the spiral-shaped structure that translates sound into nerve pulses. In normal hearing individuals, sound waves set up vibrations in the walls of the cochlea, and hair cells detect these vibrations. High frequency noises (deep notes) vibrate the base of the cochlea, while low frequency notes vibrate nearer the top of the spiral. The implant mimics the job of the hair cells. It splits the frequencies of incoming noises into a number of channels (typically eight), and then stimulates the appropriate part of the cochlea.

The two most successful implants are the Clarion (developed at the University of California at San Francisco (UCSF), Research Triangle Institute, N.C., and Advanced Bionics Corporation of Sylmar, Calif.) and the Nucleus (developed at the University of Melbourne, Australia, and made by Cochlear of Sydney, Australia). Upgrades largely focus on improving the speech processing software, which is operated by a minicomputer worn on the patient's belt. Theoretically, increasing the number of channels (and electrodes) could improve sound perception. But speech is perceived in an area of the cochlea only 14 mm long, and spacing the electrodes too close to each other causes signals to bleed from one channel to another. The result is a broad-brush version of hearing. While some recipients of the devices report speech-like sounds, many characterize their new world as being populated with legions of quacking ducks or banging garbage cans. But the success is undeniable. "Currently two thirds to three quarters of patients [with more recent models] can understand speech without lip-reading," says Steve Rebscher, a member of the UCSF team. "It's pretty amazing, and certainly better than a lot of people anticipated these devices would do." With the ear at least partially conquered, the next logical target is the eye. Several groups are working on implantable chips that mimic the action of photoreceptors, the light-sensing cells at the back of the eye. Photoreceptors are lost in both retinitis pigmentosa, a genetic disease, and age-related macular degeneration, the most common cause of lost sight in the developed world. Joseph Rizzo of the Massachusetts Eye and Ear Infirmary, and John Wyatt of the Massachusetts Institute of Technology have made a twenty-electrode, 1mm-square chip, and implanted it at the back of rabbits' eyes. The original chip, the thickness of a human hair, put too much stress on the eye, so the new version is ten times thinner.

The final set-up will include a fancy camera mounted on a pair of glasses. The camera will detect and encode the scene, then send it into the eye as a laser pulse, with the laser also providing the energy to drive the chip. Rizzo has confirmed that his tiny array of light receivers (photodiodes) can generate enough electricity to run the chip. He has also found that the amount of electricity needed to fire a nerve cell into action is ~100-fold lower in the eye than in the ear, so the currents can be smaller, and the electrodes more closely spaced. For now the power supply comes from a wire inserted directly into the eye and, using this device, Rizzo has detected signals reaching the brain. "We know we can get visual information to the brain," he says, "but we don't know what kind of vision that will result in." Eugene de Juan of Johns Hopkins Wilmer Eye Institute (Baltimore,

Md.) is trying to answer that question by using human subjects. His electrodes, inserted directly into the eye, are large and somewhat crude. But his results have been startling. Completely blind patients have seen well-defined flashes, which change in position and brightness as de Juan changes the position of the electrode or the amount of current. In his most recent experiments, patients have identified simple shapes outlined by multiple electrodes. With as little as an 8 x8 array, de Juan believes he could approximate character recognition, and a 25 x 25 array might give a crude image. The big money in eye implants is in Germany, where the government has pledged ~$10 million to two projects.

One is similar to the US projects, in which chips are implanted on the surface of the retina, the structure at the back of the eye. The other project is putting its implants at the back of the retina, where the photoreceptors are normally found. These 'subretinal' chips may block the transport of oxygen and food to the overlying nerve cells, so Eberhart Zrenner of the University of Tübingen, Germany, is developing 'chain mail' electrode arrays, with plenty of holes for the delivery of supplies.

The subretinal approach more closely mimics the action of the lost photoreceptors. The chip delivers its message at the source, before intervening layers of retinal nerve cells do some complicated processing. Rizzo thinks that his chip's hardware will be able to compensate for these computations, but for now he has other things to worry about. "There are all sorts of processes that go on that we're just going to ignore initially," he says. As with the cochlear implant, he says, "the biological tests will show you the best way to stimulate.

Ultimately the only thing that matters is what the person sees." The first big test—showing that a remotely powered device can give true visual signals—is yet to be done. "We haven't got to first base yet," says Rizzo, "but we're close." With a working device many years away, Rizzo is a model of caution. "It's always better to talk down the possibilities than to talk them up, because there are many reasons why the implants might fail," he says. "This is science fiction stuff."

3

Select Biotech Information and Capacity Building

Genomics: Selling DNA Data

Human cells are almost ridiculously tiny and efficient. Every one of them has an entire genetic instruction set—three billion As, Cs, Gs and Ts—packed into a nucleus one hundredth of a millimeter across. And a few cents worth of carbon, nitrogen and phosphorus will keep millions of cells happy in a dish, reproducing themselves and their DNA data.

Incyte Pharmaceuticals, Inc. (Palo Alto, Calif.) also stores DNA data, but in a way that is somewhat less compact and far more likely to impress a computer geek. Ground-zero for Incyte is a highly air-conditioned room filled with scores of black, towering supercomputers worth up to a million dollars each.

Nine of the ten largest pharmaceutical companies in the world each pay Incyte ~$5 million per year to look at the data on those computers. The money has brought an expanding workforce—from 160 to 675 employees in the last two and a half years—and a frantic effort to keep generating more and more information. "We're maybe a year away from going bankrupt if we don't pursue new technologies at any one point," says Tod Klingler, Director of Research Bioinformatics at Incyte.

The manpower and computer power is needed to tame an ever-growing mountain of DNA sequence information. "We now have 24 organisms completely sequenced we have their entire genetic message written down," says Temple Smith of Boston University. "This information allows us to get information about one organism and use that to understand another."

BIOCHIPS: MICROCHIPS AND MEDICINE JOIN FORCES

Paranoid ranting is a staple on the internet, and the topic of biochips is a particular favorite. Biochips—any microprocessor chips that can be used in biology—mean many things to many people. The web's preachers focus on a simple idea: an implanted chip that identifies each person uniquely, and can be used to track their location.

This wonderful idea—capable of locating lost children, downed soldiers, and wandering Alzheimers patients—sounds alarming even to many mild-mannered civil libertarians. To Living Truth Ministries of Austin, TX, it is no less than "a tremendous bonanza for the coming Antichrist!" Second Advent Ministries of San Bernardino, Calif., warns us that the US social security number will soon be replaced (at the orders of the "New World Order United Nations-controlled world government", no less) by a chip-encoded "18-digit numbering system [which] will consist of THREE separate sets of SIX DIGITS each: 6-6-6! Coincidental? No. Not if you believe Bible prophecy."

Reality is somewhat less alarming. A simple ID chip is already walking around in tens of thousands of individuals, but all of them are pets. Companies such as AVID (Norco,

Calif.), Electronic ID, Inc. (Cleburne, Tx.), and Electronic Identification Devices, Ltd. (Santa Barbara, Calif.) sell both the chips and the detectors. The chips are the size of an uncooked grain of rice, small enough to be injected under the skin using a syringe needle. They respond to a signal from the detector, held just a few feet away, by transmitting out an identification number. This number is then compared to database listings of registered pets.

Daniel Man, a plastic surgeon in private practice in Boca Raton, Fla., holds a patent on a more sinister device: a chip that would enable lost humans to be tracked by satellite. He has a large-scale model, but for miniaturization and regulatory approval he will need a substantial amount of cash. "There are some companies that are interested, but nothing is official," he says. "We're not giving up on it yet."

The civil liberties debate over biochips has obscured their more ethically benign and medically useful applications. Medical researchers have been working to integrate chips and people for many years, often plucking devices from well-known electronic appliances. Jeffrey Hausdorff of the Beth Israel Deaconess Medical Center (Boston, Mass.) has used the type of pressure-sensitive resistors found in the buttons of a microwave oven as stride timers. He places one sensor in the heel of a shoe, and one in the toe, and adds a computer to the ankle to calculate the duration of each stride.

"Young, healthy subjects can regulate the duration of each step very accurately," he says. But elderly patients prone to frequent falls have extremely variable stride times, a flag that could indicate the need for more strengthening exercises or a change in medication. Hausdorff is also using the system to determine the success of a treatment for congestive heart failure. By monitoring the numbers of strides that a person takes, he can directly measure the patient's activity level, bypassing the often flawed estimate made by the patient.

Hausdorff's chips are external, but a chip proposed by Sensors for Medicine and Science, Inc. (S4MS; Germantown, Md.) will be injected under the skin. The chip will allow diabetics to easily monitor the level of the sugar glucose in their blood. Diabetics currently use a skin prick and a hand-held blood test, and then medicate themselves with insulin depending on the result. The system is simple and works well, but the need to draw blood means that most diabetics don't test themselves as often as they should. Although they may get away with this in the short term, in later life those who monitored infrequently suffer from blindness, loss of circulation, and other complications.

The solution is more frequent testing, using a less invasive method. Various companies are trying to extract the required body fluid, using either an electric current (Cygnus, Inc., Redwood City, Calif.) or a tiny pore induced by a laser (SpectRx, Norcross, Ga.) or a small needle (Integ, St Paul, Minn.).

"All of these approaches ameliorate the drawing of blood," says George Daniloff, Chief Scientific Officer of S4MS, "but they still require manipulation of skin, and the skin may be perturbed." The S4MS chip, however, will sit underneath the skin, sense the glucose level, and send the result back out by radio-frequency communication.

A light-emitting diode (LED) starts off the detection process. The light that it produces hits a fluorescent chemical: one that absorbs incoming light and re-emits it at a longer wavelength. The longer wavelength of light is then detected, and the result is sent to a control panel outside the body.

Glucose is detected because the sugar reduces the amount of light that the fluorescent chemical re-emits. The more glucose there is, the less light that is detected. S4MS is still developing the perfect fluorescent chemical, but the key design innovation of the S4MS chip has been fully worked out. The idea is simple: the LED is sitting in a sea of the fluorescent molecules. In most detectors the light source is far away from the fluorescent molecules, and the inefficiencies that come with that mean more power and larger devices. The prototype S4MS chip uses a 22μW LED, almost forty times less powerful than the tiny power-on buttons on a computer keyboard. The low power requirements mean that energy can be supplied from the outside, by a process called induction. The fluorescent detection itself does not consume any chemicals or proteins, so the device is self sustaining.

Daniloff has a working model of an oxygen sensor that uses the same layout. With it's current circuitry it is about the size of a large shirt-button, but the final silicon wafer will be less than a millimeter square. "The scale transition is not a challenge," says Daniloff. "The challenge is to illustrate the principle." The oxygen sensor will be useful not only to monitor breathing in intensive care units, but also to check that packages of food, or containers of semiconductors stored under nitrogen gas, remain airtight.

Tom Ferrell of the Oak Ridge National Laboratory in Oak Ridge, Tenn., is also developing an oxygen-sensing chip, but his version sends light pulses out into the body. The light is absorbed to varying extents, depending on how much oxygen is being carried in the blood, and Ferrell's chip detects the light that is left. The rushes of blood pumped by the heart are also detected, so the same chip is a pulse monitor. A number of companies already make large-scale versions of such detectors, and Ferrell says he has "collaborated with almost every company in this area," although he will not name names.

Ferrell estimates that the oxygen chip is two years away, but he has already reduced the dimensions of his temperature-sensing chip to 3mm per side. The transition of certain semiconductors to their conducting state is inherently sensitive to temperature, so designing the sensor was simple enough. With some miniature radio-frequency transmitters, and foam-rubber earplugs to hold the chip in place, the device is complete. Ferrell sees applications for anyone from sick children, to chemotherapy patients who can be plagued by sudden rises in body temperature in response to their anti-cancer drugs.

Brain Surgery with an On-off Switch

Sensing and measuring is one thing, but can we switch the body on and off? Heart pacemakers use the crude approach: large jolts of electricity to synchronize the pumping of the heart. The electric pulses of the Activa implant, made by Medtronic, Inc. (Minneapolis, Minn.), are directed not at the heart but the brain. They turn off brain signals that cause the uncontrolled movements, or tremors, associated with diseases such as Parkinson's.

Drug therapy for Parkinson's disease aims to replace the brain messenger dopamine, a product of the brain cells that are dying. But eventually the drug's effects wear off, and the erratic movements come charging back. "I've had patients tell me that they scare their grandkids because they are flaying around like a fish," says Gary Heit, a neurosurgeon at Stanford University Medical Center. "Imagine your whole body twisted like a charleyhorse."

The next option is to remove a portion of the thalamus, a brain structure that triggers the tremors. The implant, cleared for use in the United States in August 1997, is a new alternative that uses high-frequency electrical pulses to reversibly shut off the thalamus. "It has the same benefits as thalamotomy [the brain operation] without the risks," says Heit. The implantation surgery is far less traumatic than thalamotomy, and if there are any post-operative problems the stimulator can simply be turned off.

Adding Back the Senses

The Activa implant interferes with aberrant brain functioning. The most ambitious bioengineers are trying to add back brain functions, restoring sight and sound where there was darkness and silence.

The success story in this field is the cochlear implant. Most hearing aids are glorified amplifiers, but the cochlear implant is for patients who have lost the hair cells that detect sound waves. For these individuals, no amount of amplification is enough.

The cochlear implant delivers electrical pulses directly to the nerve cells in the cochlea, the spiral-shaped structure that translates sound into nerve pulses. In normal hearing individuals, sound waves set up vibrations in the walls of the cochlea, and hair cells detect these vibrations. High frequency noises (deep notes) vibrate the base of the cochlea, while low frequency notes vibrate nearer the top of the spiral. The implant mimics the job of the hair cells. It splits the frequencies of incoming noises into a number of channels (typically eight), and then stimulates the appropriate part of the cochlea.

The two most successful implants are the Clarion (developed at the University of California at San Francisco (UCSF), Research Triangle Institute, N.C., and Advanced Bionics Corporation of Sylmar, Calif.) and the Nucleus (developed at the University of Melbourne, Australia, and made by Cochlear of Sydney, Australia). Upgrades largely focus on improving the speech processing software, which is operated by a minicomputer worn on the patient's belt. Theoretically, increasing the number of channels (and electrodes) could improve sound perception. But speech is perceived in an area of the cochlea only 14 mm long, and spacing the electrodes too close to each other causes signals to bleed from one channel to another.

The result is a broad-brush version of hearing. While some recipients of the devices report speech-like sounds, many characterize their new world as being populated with legions of quacking ducks or banging garbage cans. But the success is undeniable. "Currently two thirds to three quarters of patients [with more recent models] can understand speech without lip-reading," says Steve Rebscher, a member of the UCSF team. "It's pretty amazing, and certainly better than a lot of people anticipated these devices would do."

With the ear at least partially conquered, the next logical target is the eye. Several groups are working on implantable chips that mimic the action of photoreceptors, the light-sensing cells at the back of the eye. Photoreceptors are lost in both retinitis pigmentosa, a genetic disease, and age-related macular degeneration, the most common cause of lost sight in the developed world.

Joseph Rizzo of the Massachusetts Eye and Ear Infirmary, and John Wyatt of the Massachusetts Institute of Technology have made a twenty-electrode, 1mm-square chip, and implanted it at the back of rabbits' eyes. The original chip, the thickness of a human hair, put too much stress on the eye, so the new version is ten times thinner. The final set-up will include a fancy camera mounted on a pair of glasses. The camera will detect and encode the scene, then send it into the eye as a laser pulse, with the laser also providing the energy to drive the chip. Rizzo has confirmed that his tiny array of light receivers (photodiodes) can generate enough electricity to run the chip. He has also found that the amount of electricity needed to fire a nerve cell into action is ~100-fold lower in the eye than in the ear, so the currents can be smaller, and the electrodes more closely spaced.

For now the power supply comes from a wire inserted directly into the eye and, using this device, Rizzo has detected signals reaching the brain. "We know we can get visual information to the brain," he says, "but we don't know what kind of vision that will result in."

Eugene de Juan of Johns Hopkins Wilmer Eye Institute (Baltimore, Md.) is trying to answer that question by using human subjects. His electrodes, inserted directly into the eye, are large and somewhat crude. But his results have been startling. Completely blind patients have seen well-defined flashes, which change in position and brightness as de Juan changes the position of the electrode or the amount of current. In his most recent experiments, patients have identified simple shapes outlined by multiple electrodes. With as little as an 8 x8 array, de Juan believes he could approximate character recognition, and a 25 x 25 array might give a crude image.

The big money in eye implants is in Germany, where the government has pledged ~$10 million to two projects. One is similar to the US projects, in which chips are implanted on the surface of the retina, the structure at the back of the eye. The other project is putting its implants at the back of the retina, where the photoreceptors are normally found. These 'subretinal' chips may block the transport of oxygen and food to the overlying nerve cells, so Eberhart Zrenner of the University of Tübingen, Germany, is developing 'chain mail' electrode arrays, with plenty of holes for the delivery of supplies.

The subretinal approach more closely mimics the action of the lost photoreceptors. The chip delivers its message at the source, before intervening layers of retinal nerve cells do some complicated processing. Rizzo thinks that his chip's hardware will be able to compensate for these computations, but for now he has other things to worry about. "There are all sorts of processes that go on that we're just going to ignore initially," he says. As with the cochlear implant, he says, "the biological tests will show you the best way to stimulate. Ultimately the only thing that matters is what the person sees."

The first big test—showing that a remotely powered device can give true visual signals—is yet to be done. "We haven't got to first base yet," says Rizzo, "but we're close." With a working device many years away, Rizzo is a model of caution. "It's always better to talk down the possibilities than to talk them up, because there are many reasons why the implants might fail," he says. "This is science fiction stuff."

OSCP Biotechnology Team in USA

The Biotechnology Team located in EPA's Office of Science Coordination and Policy (OSCP) coordinates scientific, technical, and policy development activities within the Office of Prevention, Pesticides, and Toxic Substances (OPPTS). The OSCP Biotechnology Team is also a focal point for coordination with other Federal agencies on any issues involving biotechnology, including international activities.

BIOTECHNOLOGY CAPACITY BUILDING: A CASE STUDY

This report represents a summary of discussions that took place during a meeting held at FAO headquarters between 9th and 11th February 2005. The meeting addressed issues in connection with strengthening national plant breeding and biotechnology capacity. Over twenty representatives of agencies and institutions involved in genetic resource use and plant breeding, including an observer from the International Union for the Protection of New Varieties of Plants (UPOV) and about ten representatives from FAO attended the meeting. The participants came from approximately twenty countries.

Agriculture is the mainstay of the economies of numerous developing countries, supporting the livelihoods of large proportions of their populations. Subsistence agriculture is a particular feature of many of these countries, especially those that have suffered from the ravages of war and the vagaries of climate. With few alternatives, rural populations must often secure their sustenance and income from farming. Plant breeding, relying on judicious use of Plant Genetic Resources for Food and Agriculture (PGRFA), has supplied adapted crop varieties to many countries across the globe, with the principal exception of a number African countries. Green Revolution improved varieties have raised production levels, provided food security and generated income for many developing country farmers. The process of producing improved varieties is slow and requires long-term sustained commitment to plant breeding. In many countries application of biotechnology to agricultural research and production has hailed a new era. There is the promise that application of biotechnology can speed up some aspects of plant breeding, although the techniques complement rather than substitute for conventional plant breeding methods. However, neither the potential of biotechnology nor the proven value of plant breeding in general can be realised in the absence of adequate investment. Unfortunately public sector investment in plant breeding and biotechnology, and the research that supports them, has been in decline over recent times. National Agricultural Research Systems (NARS) have been the principal sufferers in this resource-constrained environment as plant breeding programmes have been trimmed and the possibilities for introducing biotechnological tools have receded. Private sector investment in agricultural research and plant breeding, often using the tools of biotechnology, has increased, but linkages with the public sector have been sub-optimal. Furthermore, scarce attention is paid to crops where seed trade is minimal, and which are often important crops of smallholder agriculture in developing countries. It has been suggested that realisation of the Millennium Development Goals might be facilitated by increased investment in plant breeding in both developed and developing nations. The declining capacity of many national plant breeding programmes has also meant that fewer crops can be researched and bred. There are many vitally important crops that have not received adequate attention from scientists and plant breeders. Given the current financial constraints, it seems unlikely they will receive the resources they merit to enable diversification for food security and poverty alleviation. The International Treaty (IT) on PGRFA, which has been ratified by 66 countries, serves to enhance access to and sustainable use of PGRFA. The Global Crop Diversity Fund was established to contribute to PGRFA conservation efforts. The issue remains, however, that conservation only represents half of the problem. While PGRFA conservation is crucial for providing the raw material for plant breeders to work with, PGRFA use is key to sustaining livelihoods, ensuring food security and contributing to national development.

The national plant breeding survey questionnaire

To quantify the status and progress in investment and production of national plant breeding programmes, FAO considered it necessary to implement a survey. The approach to the issue in general, including the questionnaire in particular, was designed through an expert consultation. This in turn led to the paper Towards the Sustainable Use of Plant Genetic Resources Food and Agriculture (PGRFA): Strengthening Plant Breeding, which was prepared for the Working Group on PGRFA of the CGRFA (CGRFA/WG-PGR-2/03/2). The questionnaire is included. It is being used to gather information on plant breeding capacity in terms of budget allocation and staff establishment between 1985 and 2004. Moreover, it requires information to be supplied on the crops addressed in national programmes and the extent of application of biotechnology to plant breeding. At the time of the workshop, 18 countries had completed the questionnaire. It is planned to build on the results of the pilot phase and send the questionnaire to many more countries, with the ultimate aim of producing and maintaining an up-to-date database on national plant breeding programmes. Some of the information from the completed questionnaires was presented and discussed at the workshop.

Objectives and structure of the workshop

The objectives of this workshop were to bring together participants from developing countries of Africa, Asia and South America, representing a variety of backgrounds and expertise, to discuss the problem of the decline in plant breeding and how this might be remedied. The workshop aimed to provide guidance on, and detail linkages between, modern and traditional plant breeding. The necessary move from conservation of PGRFA to sustainable use was also a major discussion point.

The workshop programme is given a list of the workshop participants is provided. The workshop comprised presentations of country profiles, based on the results of the questionnaire. Discussion groups addressed many of the principal issues and constraints concerning plant breeding status and direction in developing countries. The presentations and discussions were synthesised into the following text and recommendations.

Preamble—A complex issue

Development of national plant breeding capacity and its current and future status is a complex issue. Geographical and economic circumstances differ enormously over the developed and developing world. Even on a regional basis, there can be large differences among countries. Civil conflict and natural disasters have had long-lasting effects on national capacities to undertake agricultural research. Countries also have very different natural resource bases; some are rich and others poor. Some have well-educated workforces while others do not. Given such complicating factors, it is not possible to describe a single model of the status of plant breeding and biotechnology application in the developing world. Nor is it possible to design a single set of remedial measures needed to halt the decline in plant breeding. Accurate models will necessarily be situation-specific.

Working group discussion

Presentations and discussions took place over the course of three days. For convenience, the main ideas and issues are discussed under six broad headings.

Public and private sectors: The public sector, in particular the NARS, has the largest responsibility for conducting agricultural research and applying the results. This includes producing improved crop varieties for cultivation by a country's farmers. The capacity to undertake this work varies enormously among countries and even within regions. This was highlighted during the presentations of country data and regional syntheses. While some countries have taken on board some of the more advanced methods of biotechnology, most are struggling to maintain, rehabilitate and initiate meaningful plant breeding programmes. Although the overall trend is towards reduced funding and decreased capacity to produce new, adapted varieties of essential crops, there are certain regions that are able to do a better job than others are.

A few countries are very advanced. One of the principal factors underlying the problem of reduced funding faced by the public sector is lack of awareness among policy makers of the impact of plant breeding on national development. Those countries that were not able to benefit from the improved varieties and farming practices that derived from the Green Revolution have not made as much progress as those that were able to benefit. It is generally under-appreciated that the returns to investment in plant breeding are substantial and that without plant breeding it is difficult to effect progress in agrarian societies.

The private sector encompasses a range of enterprises, including plant breeders, processing industries, seed distributors, agrochemical producers and marketers among others. Numerous private sector companies are integral components of the agricultural economy of many countries, where they play essential roles. During recent times, some companies in the private sector, particularly large multinationals have invested heavily in agricultural research, and particularly in biotechnology. The results of much of this research are not publicly available and cannot be used by national programmes to improve the situation of small farmers in developing countries. In the absence of regulatory frameworks in many developing countries, adapted crop varieties that incorporate elements of genetic engineering cannot be deployed. Moreover, private industry has concentrated heavily on a small number of crops of global importance. These are not necessarily the most important crops for small-scale farmers who often have to meet subsistence needs before they can consider cultivating crops to generate income.

The major issue in connection with the roles of the public and private sector is that they largely operate independently. In addition, the private sector is often able to attract the best qualified and most able individuals, to the detriment of the public sector. Public-private joint ventures, based on complementarity rather than competition, were seen by the working group as being vital to supporting agricultural production and plant breeding in developing countries. Co-operatives were considered a good example of how small-scale farmers could benefit from a more business-like approach to crop production and retail, especially in the seed sector. Encouraging small-scale agricultural enterprises was seen as a good way to boost development.

Donors and funding: With changing priorities among multilateral, bilateral and national donors, there has been a general move away from funding plant breeding. Not only do many donors have agendas that are not set according to recipient country needs, the agendas also change relatively rapidly in comparison with the long-term commitment that is necessary to realise benefits from plant breeding. A further marked failing is the fragmented way in which donors often operate. A coordinated, longer-term response was considered necessary to be able to achieve results in plant breeding that could benefit farmers and consumers in the developing world. Not only is it imperative that plant breeding programmes are funded over the long term, as plant breeders are not able to work without operational budgets, but capacity building also has to be addressed to make sure that there are trained personnel available to do the work.

Capacity building: Selected results from the questionnaire and consequent discussion indicated that declining plant breeding capacity, in terms of stagnant or reduced budgets and fewer released varieties, was confounded in some instances by increases in staff establishment and qualifications. One factor underlying this was changing age profiles of staff. Young, wellqualified people are attracted to other professions that offer better conditions and career prospects. Moreover, although biotechnology might feature among the more attractive subjects, it is very often isolated from practical plant breeding. What is required currently is a workforce that can use the tools of biotechnology in plant breeding. Without adequate numbers of suitably qualified staff, productive plant breeding will not be possible. Capacity building to achieve these ends was considered a priority area. The international agricultural research centers of the CGIAR have tremendous potential for contributing to capacity building. Not only do they conduct research on many of the world's most important crops, but they also represent an opportunity for NARS to send staff for training and retraining in a wide variety of subjects. Furthermore, ample field, laboratory and greenhouse facilities allow practical work to be pursued in cooperation with academic institutions elsewhere, including regional centers of excellence, which might provide training in plant breeding theory. Several CGIAR centres contribute substantially to NARS plant breeding efforts through supplying advanced germplasm, but they do not substitute for strong NARS: they strengthen NARS.

Regulatory issues: Plant breeders' rights (PBR) and intellectual property rights (IPR) were discussed as a means of encouraging private-public cooperation through promotion of public sector income generation, while protecting the interests of breeders. They are also able to accommodate participatory plant breeding. Subsistence farmers1, moreover, are exempt from much of the legislation involved in PBR, and are able to save and sow seed of protected varieties without the need for permission of the breeder. It was considered that PBR suited a variety of circumstances, although not all.

Prospects for regional cooperation in plant breeding were discussed. Where crops and cropping systems are common to a region it might be possible to mount a regional approach to plant breeding, and thereby reduce overall costs and limit unnecessary and unproductive duplication and competition. This will only be feasible under circumstances where associated legislation, particularly that regarding transfer of plant material across national borders, is harmonised. An added benefit of this approach might be that the stronger NARS could support weaker ones in a region.

Other forms of novel policy implementation to promote funding were discussed, including levies on products (not necessarily agricultural products) that could be fed back into the system to support research.

Biotechnology: Biotechnology is an integral feature of modern plant breeding research and practice. It is not however sufficiently well embedded in practical plant breeding in developing countries. It covers a wide range of methods, from relatively straightforward and cheap tissue culture, through marker assisted selection to genetic engineering. Some, and in a few cases all, of the biotechnologies associated with crop development have become part of plant breeding programmes in developing countries. Inadequate infrastructure, inadequate funding and lack of trained staff represent major obstacles to uptake however. These issues have to be addressed if plant breeding in developing countries is to be kept up-to-date. Capacity building in plant biotechnology will be most beneficial when staff are part of or closely linked with plant breeding programmes.

Next steps: It was proposed, and generally accepted, that there was a need for more consultations before any decisions could be reached on the way forward for strengthening national plant breeding and biotechnology capacity. It was suggested that a workshop should be organized with country representatives and decision-makers to increase awareness of the benefits of plant breeding and application of appropriate biotechnology to plant breeding.

Presenting work on small-scale seed systems and establishing the links with plant breeding could follow this. Engagement with the private sector would be beneficial to establish common ground. Finally, having developed a strategy for strengthening national plant breeding capacities, meetings could be held with donors to discuss future funding. It was thought necessary to keep these steps separate and move relatively slowly rather than attempt to address all issues and include numerous stakeholders at a single meeting.

It was agreed that it would be useful to move forward with the national plant breeding survey, including more countries. There was a perceived need to address some of the issues concerning what has happened to plant breeding staff and what underlies the observed trends in staffing.

It is essential to complete analysis of the data already received and publish it. Data up to 2004 have been requested and it was suggested that information on privatepublic linkages would represent a useful addition to those data already collected. The survey should be reviewed regularly by the participating countries.

The possibilities of setting up a special fund were discussed, with specific reference to linking with the private sector, and particularly for crops that do not currently interest the private sector. The IT was considered a possible mechanism to leverage funds to support small-scale seed enterprises with plant breeding activities. During the course of the discussion, it was suggested that FAO might consider publishing a text describing success stories in development of small-scale agro-industries, seed industries in particular, and the linkages with plant breeding.

Recommendations arising from the workshop

The following recommendations were formulated following three days of discussion by the working group on the major issues and problems associated with plant breeding and application of biotechnology in the public sector in developing nations. It was recognised that both rural and urban societies, and national economies of the developing world, are largely reliant on the agricultural sector. Nations with weak agricultural sectors, and particularly those without or with only poor plant breeding support, are at a disadvantage in comparison with those that are able to provide their farmers with improved crop varieties and appropriate technologies. Strengthening infrastructure and building capacity represent cornerstones of policies aimed at addressing current shortcomings. Strengthening sustainable agriculture through plant breeding and the application of appropriate biotechnology is of primary importance. The recommendations made by the working group were differentiated on the basis of major stakeholder groups.

Recommendations to FAO

1. Increase awareness among all stakeholders of the strategic importance of sustainable crop improvement through plant breeding.

2. Encourage all stakeholders in plant breeding in developing and developed countries to increase awareness of the value and impact of plant breeding and application of appropriate biotechnology on development.

3. Take note of the momentum that has been created through the Convention on Biological Diversity, the IT on PGRFA and the Global Crop Diversity Fund, and establish a platform for strategic thinking in a systematic manner for future development of sustainable use of PGRFA. This would necessarily include plant breeding and the application of biotechnology.

4. Consider establishing a fund to support sustainable plant breeding, for crops that have not received due attention of plant breeders and crop scientists, including genetic enhancement, in countries and for programmes requiring support. Capacity building would be a key component.

Recommendations to the Commission on Genetic Resources for Food and Agriculture (CGRFA) and the governing body of the International Treaty on PGRFA

1. Promote discussion of the use of PGRFA through plant breeding as an essential extension to conservation efforts. The IT might assist in strengthening the mechanisms and modalities for capacity building and technology transfer to developing countries (article 13 of the IT) to enable them to realise the potential of the PGR of their own farmers' varieties and the PGR made available through the multilateral system of the IT.

2. Encourage PGRFA users, particularly plant breeders, to contribute to CGRFA discussions and decision-making on sustainable use to complement efforts on conservation. Member countries of the CGRFA, where possible, should consider including specialists in sustainable use of PGRFA in their delegations to meetings of the Commission and IT Governing Body.

3. Give urgent consideration to capacity building and technology transfer to developing countries so that they are able to use PGRFA in plant breeding programmes and contribute to the sustainable development of agriculture and allied activities.

4. Consider establishing a fund to support sustainable plant breeding for crops that have not received the due attention of plant breeders and crop scientists, including genetic enhancement, in countries and for programmes requiring support. Capacity building would be a key component.

Recommendations to NARS

1. Improve priority setting, incorporating awareness of markets and consumption chains and the paths of scientific and technological innovation.

2. Play an increasingly active role in decision making, especially in connection with capacity building and the need to strengthen it.

3. Adopt a multi-disciplinary approach to plant breeding, including biotechnology where appropriate.

4. Attempt to develop a solid and stable infrastructural and financial national base to ensure sustainability of crop improvement.

5. Seek to develop improved linkages, including those with farmers and the private sector, particularly with the informal seed sector where possible.

6. Increase awareness of the strategic importance of sustainable crop improvement.

7. Monitor and evaluate the impacts of IPR/IP on the development and effectiveness of breeding so that adjustments to strategies can be made on an informed basis.

Recommendations to Donors

1. Listen to all parties and promote awareness with a view to supporting plant breeding and its enhancement in developing countries.

2. Exert influence in developing national bases of support.

3. Improve coordination among donors.

4. Support networks to enhance capacity of national programmes, including through mechanisms such as the facilitating mechanism of the Global Plan of Action.

5. Support efforts to facilitate merging and integration of conventional and advanced breeding strategies.

6. Contribute to initiatives that foster sustained funding and long-term support needed to carry out plant breeding.

Recommendations to the CGIAR

1. Review survey documentation and advise on interpretation and strategic investments in crop improvement.

2. Balance donor demands and the needs of NARS and continue to support plant breeding through conducting research and capacity building according to the needs of

national programmes and national programme staff.

3. Monitor the impact of capacity building efforts with a view to maintaining focused programmes.

4. Encourage and support cooperation at regional and sub-regional levels, and enhance south-south linkages.

5. Coordinate efforts to facilitate merging and integration of conventional and advanced breeding strategies.

6. Providc leadership in transfer of tools and technologies to NARS.

Recommendations to Policy-makers

1. Recognise the continued importance of agriculture to the national economies of developing nations and to the livelihoods of the rural and urban populations therein.

2. Recognise the contribution of plant breeding to improved food security and income generation in developing nations.

3. Understand the long-term nature of plant breeding and the needs for sustained support to ensure its success.

4. Encourage multi-stakeholder participation in decision-making processes that support conservation and use of PGRFA.

5. Encourage highly professional debate, relying on specialised representation, in forums such as the IT, on plant breeding and stress its importance in maintaining livelihoods in developing nations.

6. Harmonise policies related to sustainable use of PGRFA including seed systems.

7. Examine and understand the linkages between sustainable use and conservation of PGRFA.

8. Assist in implementing appropriate forms of rights for the developers and users of plant varieties used in agriculture in developing nations.

9. Monitor and evaluate the impacts of IPR/IP on the development and effectiveness of breeding so that adjustments to strategies can be made on an informed basis.

Agricultural Biotechnology

Agricultural biotechnology includes, but is not limited to:

- using our knowledge of cells to modify their activities.
- modifying cells or microorganisms using the techniques of recombinant DNA rather than traditional selective breeding.
- producing commercial quantities of desirable substances using microorganisms, cells or parts of cells.

Some examples of techniques used in agricultural biotechnology are:

- selective breeding that produces crops and livestock with specified characteristics.
- utilizing chemicals such as colchicine or radiation to induce mutations in genetic material.
- growing new plants using specific parts of another plant in a process called vegetative propagation. Grafting is one example that is important in fruit trees.
- encouraging the growth of a new plant in a prepared nutrient medium within a controlled environment.

Current examples of agricultural biotechnology products on the market are:

- corn plants having resistance to a particular herbicide used to kill weeds.
- corn plants producing a protein toxic to insect pests reducing the amount of insecticide use.
- tomatoes producing less of a hormone that causes tomatoes to ripen thereby extending the shelf life.
- corn hybrid plants growing in strongly alkaline soils that are found in the western US corn belt.
- cotton plants resisting cotton bollworms and tobacco budworms.

A large percentage of watermelons imported into the US annually come from Mexico with the remaining volume arriving from Central and South America. Imports have increased to meet the consumer demand for watermelon all-year-round.

REFERENCES

Andres, V. and Walsh, K. (1996) Myogenin expression, cell cycle withdrawal and phenotypic differentiation are temporally separable events that precedes cell fusion upon myogenesis. J. Cell Biol., 132, 657-666.

Asahara T, Takahashi T, Masuda H, Kalka C, Chen D, Iwaguro H, Inai Y, Silver M, Isner JM. VEGF contributes to postnatal neovascularization by mobilizing bone marrow-derived endothelial progenitor cells. EMBO J 1999; 18:3964- 3972.

Beltrami AP, et al. Evidence that human cardiac myocytes divide after myocardial infarction. *N Engl J Med* 344, 1750-7 (2001).

BZ Atkins, MT Hueman, JM Meuchel, MJ Cottman, KA Hutcheson, DA Taylor: Myogenic cell transplantation improves in vivo regional performance in infarcted rabbit myocardium. J Heart Lung Transpl 1999, 18: 1173-1180.

CE Murry, RW Wiseman, SM Schwartz, SD Hauschka: Skeletal myoblast transplantation for repair of myocardial necrosis. J Clin Invest 1996, 98: 2512-2523.

Choi K, Kennedy M, Kazarov A, Papadimitriou, Keller G. A common precursor for hematopoietic and endothelial cells. Development 1998;125:725-732.

Colucci WS. Molecular and cellular mechanisms of myocardial failure. Am J Cardiol 1997;80:15L-25L.

DA Taylor, BZ Atkins, P Hungspreugs, TR Jones, MC Reedy, KA Hutcheson, DD Glower, WE. Regenerating functional myocardium: improved performance after skeletal myoblast transplantation. Nature Medicine 1998, 4: 929-933.

Effects of enalapril on mortality in severe congestive heart failure. Results of the Cooperative North Scandinavian Enalapril Survival Study (CONSENSUS) Trial Study Group. N Engl J Med 1987; 316(23):1429-1435.

Elefanty AG, Robb L, Birner R, Begley CG. Hematopoietic-specific genes are not induced during in vitro differentiation of scl-null embryonic stem cells.

Folkman, J. Therapeutic angiogenesis in ischemic limbs. Circulation 1998; 97:108- 110.

Frazier OH, Rose EA, Macmanus Q, et al. Multicenter clinical evaluation of the Heartmate 1000 IP left ventricular assist device. Ann Thorac Surg 1992; 102:578-587.

Halevy, O., Novitch, B.G., Spicer, D.B., Skapek, S.X., Rhee, J., Hannon, G.J., Beach, D. and Lassar, A.B. (1995) Correlation of terminal cell cycle arrest of skeletal muscle with induction of p21 by MyoD. Science, 267, 1018- 1021.

Hescheler J, Fleischmann BK, Wartenberg M, Bloch W, Kolossov E, Ji G, Addicks K, Sauer H (1999) Establishment of ionic channels and signalling cascades in the embryonic stem cell-derived primitive endoderm and cardiovascular system. Cells Tissues Organs 165:153-164.

Heymans S, Luutun A, Nuyens D, et al. Inhibition of plasminogen activators or matrix metalloproteinases prevents cardiac rupture but impairs therapeutic angiogenesis and causes cardiac failure. Nat Med 1999;5:1135- 1142.

Hill MF, Singal PK. Right and left myocardial antioxidant responses during heart failure subsequent to myocardial infarction. Circulation 1997 96:2414-20.

Hochman JS and Choo H. Limitation of myocardial infarct expansion by reperfusion independent of myocardial salvage. Circulation 1987;75:299- 306.

Hodgetts SI, Beilharz MW, Scalzo T, Grounds MD (2000) Why do cultured transplanted myoblasts die in vivo? DNA quantification shows enhanced survival of donor male myoblasts in host mice depleted of CD4+ and CD8+ or NK1.1+ cells. Cell Transplant 9:489-502.

Hognes JR. In Thc Artificial Heart: Prototypes, Policies and Patients. Washingtion, DC: National Acadcmy Press; 1991:1-312.

Jackson KA, Majka SM, Wang H, Pocius J, Hartley CJ, Majesky MW, Entman ML, Michael LH, Hirschi KK, Goodell MA (2001) Regeneration of ischemic cardiac muscle and vascular endothelium by adult stem cells. J Clin Invest 107:1395-1402.

Jaffredo T, Gautier R, Eichmann A, Dieterlen-Lievre F. Intraaortic hemopoietic cells are derived from endothelial cells during ontogeny. Development 1998;125:4575-4583.

Kajstura J, Leri A, Finato N, di Loreto N, Beltramo CA, Anversa P. Myocyte proliferation in end-stage cardiac failure in humans. Proc Natl Acad Sci USA 95, 8801-8805 (1998).

Kalka, C. et al. Transplantation of ex vivo expanded endothelial progenitor cells for therapeutic neovascularization. Proc. Natl. Acad. Sci. USA 2000; 97:3422-3427.

Kalkman EAJ, Bilgin YM, van Haren P, van Suylen R-J, Saxena PR, Schoemaker RG.Determinants of coronary reserve in rats subjected to coronary artery ligation or aortic banding. Cardiovasc Res 1996.

Kawano H, Do YS, Kawano Y, Starnes V, Barr M, Law RE, Hsueh WA. Angiotensin II has multiple profibrotic effects in human cardiac fibroblasts. Circulation 2000;101:1130-1137.

Kehat I, Kenyagin-Karsenti D, Snir M, Segev H, Amit M, Gepstein A, Livne E, Binah O, Itskovitz-Eldor J, Gepstein L. Human embryonic stem cells can differentiate into myocytes with structural and functional properties of cardiomyocytes. J Clin Invest. 2001; 108: 407-14.

Kellerman S, Moore JA, Zierhut W, Zimmer HG, Campbell J, Gerdes

AM. Nuclear DNA content and nucleation patterns in rat cardiac myocytes from different models of cardiac hypertrophy. *J Mol Cell Cardiol* , 497-505 (1992).

Kennedy M, Firpo M, Choi K, Wall C, Robertson S, Kabrun N, Keller G. A common precursor for primitive erythropoiesis and definitive haematopoiesis.

Klug MG, Soonpaa MH, Koh GY, Field LJ (1996) Genetically selected cardiomyocytes from differentiating embryonic stem cells form stable intracardiac grafts. J Clin Invest 98:216-224.

Kocher A, Schuster M, Szabolcs M, Itescu S. Cardiomyocyte regeneration after neovascularization of ischemic myocardium by human bone marrowderived angioblasts. Nature Medicine 2002, in press.

Kocher AA, Schuster MD, Szabolcs MJ, Takuma S, Burkhoff D, Wang J, Homma S, Edwards NM, Itescu S. Neovascularization of ischemic myocardium by human bone-marrow-derived angioblasts prevents cardiomyocyte apoptosis, reduces remodeling and improves cardiac function. Nature Med.

L Field: Future therapy for cardiovascular disease. In Proceedings of the NHLBI Workshop Cell Transplantation: Future Therapy for Cardiovascular Disease? Columbia, MD; 1998.

Labastie M-C, Cortes F, Romeo P-H, Dulac C, Peault B. Molecular identity of hematopoietic precursor cells emerging in the human embryo. Blood 1998;92:3624-3635.

Lancet 347:279.

Levkau B, Koyama H, Raines EW, Clurman BE, Herren B, Orth K, Roberts JM, Ross R. Cleavage of p21cip1/waf1 and p27 kip1 mediates apoptosis in endothelial cells through activation of cdk2: role of a caspase cascade.

Li, Y., Jenkins, C.W., Nichols, M.A. and Xiong, Y. (1994) Cell cycle expression and p53 regulation of the cyclin-dependent kinase inhibitor p21. Oncogene, 9, 2261-2268 Steinman, R.A., Hoffman, B., Iro, A., Guillouf, C., Liebermann, D.A. and El- Houseini, M.E. (1994) Induction of p21 (WAF1/CIP1) during differentiation.

Liechty KW, MacKenzie TC, Shaaban AF, et al. Human mesenchymal stem cells engraft and demonstrate site-specific differentiation after in utero transplantation in sheep. Nat Med 2000, 6:1282-6.

McCarthy PM, Rose EA, Macmanus Q, et al. Clinical experience with the Novacor ventricular assist system. J Thorac Cardiovasc Surg. 1991; 102:578-587.

McEwan PE, Gray GA, Sherry L, Webb DJ, Kenyon CJ. Differential effects of angiotensin II on cardiac cell proliferation and intramyocardial perivascular fibrosis in vivo. Circulation 1998;98:2765-2773.

Nelissen-Vrancken H, Debets J, Snoeckx L, Daemen M, Smits J. Time-related normalization of maximal coronary flow in isolated perfused hearts of rats with myocardial infarction. Circulation 1996;93:349-355.

Nidorf SM, Siu SC, Galambos G, Weyman AE, Picard MH. Benefit of late coronary reperfusion on ventricular morphology and function after myocardial infarction. J Am Coll Cardiol 1992;20:307-313.

Ogawa M, Kizumoto M, Nishikawa S, Fujimoto T, Kodama H, Nishikawa SI. Blood 1999;93:1168-1177.

Pitt B, Segal R, Martinez FA, et al. Randomised trial of losartan versus captopril in patients over 65 with heart failure (Evaluation of Losartan in the Elderly Study, ELITE). Lancet 1997;349:747-752.

Pittenger MF, Mackay AM, Beck SC, et al. Multilineage potential of adult human mesenchymal stem cells. Science 1999, 284:143-147.

Pouzet B, Ghostine S, Vilquin J-T, Garcin I, Scorsin M, Hagege AA, Duboc D, Schwartz K, Menasche P (2001) Is skeletal myoblast transplantation clinically relevant in the era of angiotensin-converting enzyme inhibitors? Circulation 104:I223-228.

Rafii S, Shapiro F, Rimarachin J, Nachman R, Ferris B, Weksler B, Moore AS, Asch AS. Isolation and characterization of human bone marrow microvascular endothelial cells: hematopoietic progenitor cell adhesion.

Ravichandran LV and Puvanakrishnan R. In vivo labeling studies on the biosynthesis and degradation of collagen in experimental myocardial myocardial infarction. Biochem Intl 1991;24:405-414.

Shi Q, Rafii S, Wu MH-D, et al. Evidence for circulating bone marrow-derived endothelial cells. Blood 1998;92:362-367.

Smythe GM, Grounds MD (2000) Exposure to tissue culture conditions can adversely affect myoblast behaviour in vivo in whole muscle grafts: implications for myoblast transfer therapy. Cell Transplant 9:379-393.

Smythe GM, Hodgett's SI, Grounds MD (2001) Problems and solutions in myoblast transfer therapy. J Cell Mol Med 5:33-47.

*Takahashi, T. et al. Ischemia- and cytokine-induced mobilization of bone marrowderived endothelial progenitor cells for neovascularization. Nat. Med. 1999; 5:434-438.

Tavian M, Coulombel L, Luton D, San Clemente H, Dieterlen-Lievre F, Peault B.

4

Sample Academic Programs in Biotechnology

DEGREE PROGRAMS IN BIOCHEMISTRY AND BIOTECHNOLOGY

General Information

Degrees

The Biochemistry and Biotechnology Program provides academic programs leading to the undergraduate B.S. degree or the graduate M.S. degree in Biochemistry and Biotechnology. These degrees are offered in cooperation with the Department of Biology and the Department of Chemistry and Biochemistry. Faculty members in this program are engaged in teaching and research in areas such as biochemistry, genetics, molecular biology, cell biology and developmental biology. Majors have the opportunity through coursework, laboratories, seminars and research experience to develop the knowledge and skills necessary to enter the workforce or to go on with further graduate education.

Career Outlook

Nationally and regionally the emerging Biotechnology sector will increase the demand for workers with significant training in molecular biology, biochemistry, and genetics.

The St. Louis metropolitan area has long been a major center for biochemistry and biotechnology, and in the past decade it has become a focus for the establishment of life sciences research and development. A degree in Biochemistry and Biotechnology provides students with the training they need to become part of the broad biotechnology and life sciences industries.

Undergraduate Studies

Degree Requirements: Bachelor of Science in Biochemistry and Biotechnology

General Education Requirements

Students must satisfy the university and college general education requirements. Some math or science courses required for the major may be used to meet the science and mathematics requirement of the university. There is no foreign language requirement for the degree.

Satisfactory/Unsatisfactory Option

Up to 18 credit hours may be taken on a satisfactory / unsatisfactory (s/u) basic. Excluded from this option are required courses in biology, chemistry, physics, and mathematics.

Non-major Biology or Chemistry Courses

Courses in Biology with a number less than 1800 and courses in Chemistry with a number less than 1100 do not count toward the credit hours required for a major in biochemistry and biotechnology.

1. Biology and Chemistry Core Courses

   Biol 1811, Introductory Biology I

   Biol 2012, Genetics

   Biol 2013, Genetics Laboratory

   Biol 2482, Microbiology

Biol 2483, Microbiology Laboratory

Biol 3622, Cell Biology

Biol 4602, Molecular Biology

Chem 1111, Introductory Chemistry I

Chem 1121, Introductory Chemistry II

Chem 2223, Quantitative Analysis

Chem 2612, Organic Chemistry I

Chem 2622, Organic Chemistry II

Chem 2633, Organic Chemistry Laboratory

Chem 3302, Physical Chemistry for Life Sciences

2. Biochemistry and Biotechnology Core Courses

Biol 4712 or Chem 4712, Biochemistry

Biol 4713, Techniques in Biochemistry or

Chem 4733, Biochemistry Laboratory

Biol 4614, Biotechnology Laboratory I

3. Math and Physics Core Courses

Math 1030, College Algebra

Math 1035, Trigonometry

Math 1100, Basic Calculus or

Math 1800, Analytic Geometry and Calculus I

Phys 1011, Basic Physics

Phys 1012, Basic Physics

4. Elective Courses – 9 credit hours chosen from the courses below:

Biol 4612, Molecular Genetics of Bacteria

Biol 4632, Nucleic Acid Structure and Function

Biol 4622, Molecular Cell Biology

Biol 4615, Biotechnology Laboratory II

Chem 4722, Advanced Biochemistry

Chem 4764, Interdisciplinary Topics in Biochemistry

Chem 4772, Physical Biochemistry

Electives

Recommendations include basic statistics (Math 1310 or Math 1320), computer science, public speaking (Comm 1040), foreign language, ethics, and undergraduate research.

Research Opportunity

Students are encouraged to complete a minimum of 2 credit hours of undergraduate research, which provides students with an opportunity to gain research experience under the supervision of a faculty member. The project will normally include a review of the literature, laboratory experience and a research paper.

Minor in Biology

Students complete the B.S. degree in Biochemistry and Biotechnology may also obtain a minor biology by completing Biology 1821, Introductory Biology. The minor must be approved by the chairperson of the Department of Biology. At least 9 hours of the biology course credits must be taken in residence at UM-St. Louis. Candidates must have a cumulative grade point average of 2.0 or better in the minor, and none of the courses may be taken on a satisfactory/unsatisfactory (s/u) basis.

Minor in Chemistry

Students complete the B.S. degree in Biochemistry and Biotechnology will also fulfill the course requirements for a minor in Chemistry. A GPA of at least 2.0 is required for the courses presented for the minor. At least three courses toward the Chemistry minor must be completed at UM-St. Louis.

Graduate Studies

Degree Requirements

Master of Science in Biochemistry and Biotechnology

The Biochemistry and Biotechnology Program offers two types of Master of Science degrees. One is a non-thesis option suitable for those with laboratory research experience or for others, such as educators, who do not require research experience. The other option includes a laboratory-based research apprenticeship leading to a written thesis. All students admitted to the graduate program are considered to be in the non-thesis program unless they have been accepted as a thesis student by a faculty member of the program.

M.S. Admission Requirements

Applicants to the M.S. program must submit completed application and personal data forms, three letter of recommendation from undergraduate faculty, and transcripts of all previous work. Submission of Graduate Record Examination scores, although not required, is highly recommended and will be used by the admission committee for consideration of admittance. Admission as a regular graduate student requires graduation from an accredited college with a minimum grade point average overall and in biology and chemistry courses of 3.0 (where A = 4.0). Students will generally be expected to have completed a major in biology, chemistry, or biochemistry. In addition to the Graduate School admission requirements, applicants should have completed an advanced undergraduate course in biochemistry.

All foreign applicants, except those from countries where English is the primary language, must take the TOEFL. Ordinarily, a score of 213 on the computer-based exam (550 on the paper-based exam) or better is required.

Requirements

Both the thesis and non-thesis options require a total of 30 graduate credit hours, of which at least half must be at the 5000-level or above. A maximum of 12 or 5 credit hours of Graduate Research (Biol or Chem 6905) may be applied toward the 30 credit hour total for the thesis or non-thesis options, respectively.

1. Required Courses

   Chem 4722, Advanced Biochemistry

   Biol 4614, Biotechnology Laboratory I or

   Biol 6615, Advanced Biotechnology Laboratory II

   Biol 6602, Advanced Molecular Biology or

   Biol 6612, Advanced Molecular Genetics of Bacteria

   Biol 6889, Graduate Seminar

   Chem 5794, Special Topics in Biochemistry (Bioinformatics)

2. Elective Courses

   Chem 4733, Biochemistry Laboratory

   Chem 4772, Physical Biochemistry

   Chem 4764, Interdisciplinary Topics in Biochemistry

   Chem 5794, Special Topics in Biochemistry

   Chem 6787, Problem Seminar in Biochemistry

   Chem 6905, Graduate Research

   Biol 4842, Immunobiology

   Biol 5069, Topics in Cell and Molecular Biology

   Biol 6602, Advanced Molecular Biology

   Biol 6612, Advanced Molecular Genetics of Bacteria

   Biol 6615, Advanced Biotechnology Laboratory II

   Biol 6622, Advanced Molecular Cell Biology

Biol 6632, Advanced Nucleic Acid Structure and Function

Biol 6642, Advanced Plant Molecular Biology & Genetic Engineerng

Biol 6652, Advanced Virology

Biol 6699, Graduate Internship in Biotechnology

Biol 6889, Graduate Seminar

Biol 6905, Graduate Research

Course Descriptions

Prerequisites may be waived by consent of the department or instructor. Some courses as indicated in the course description may be taken concurrently with the listed offering. Consult an adviser for further information.

Students who have earned 24 or more semester hours of credit at any accredited post-secondary institution(s) before the start of the fall 2002 semester must meet the general education requirements stipulated in the UM-St. Louis 2001-2002 Bulletin. The following courses fulfill the Natural Sciences and Mathematics breadth of study requirements as described in that Bulletin:

Biology: 1811, 2012, 2482, 3622, 4602, 4712, 4612, 4622, 4632. Chemistry: 1111, 1121, 2223, 2612, 2622, 3302, 4712, 4722, 4764, 4772.

Biology Courses

*Note – Majors in Biochemistry and Biotechnology are not required to take Biol 1821; hence it is not a prerequisite for Biol 2012, Biol 2482, or Biol 3622 for these majors.

1811 Introductory Biology: From Molecules to Organisms (5), [MI, MS]

Prerequisite: A minimum of high school chemistry,

English 1100 or equivalent (may be taken concurrently), and placement into college algebra or higher. Required for students intending to major in biology or take specified biology courses at the 2000-level or above. This course presents an introduction to some of the principles of biology and scientific methodology applied to the molecular/cellular through organ system levels of organization. Topics include: cell structure, metabolism, reproduction, heredity and major physiological processes regulated by organ systems. Three hours of lecture, three and one-half hours of lab, and one hour of discussion per week.

2012 Genetics (3)

Prerequisite: Biology 1811[majors must also take Biology 1821] and Chem 1111 or [Chem 1082 plus Chem 1091]. Fundamental principles of inheritance, including classical genetic theory as well as recent advances in the molecular basis of heredity. Three hours of lecture per week.

2013 Genetics Laboratory (2)

Prerequisite: Concurrent registration in Biology 2012, or by consent of instructor. Laboratory to accompany Biology 2012. Three and one-half hours of organized laboratory time per week. Students may need to return to the laboratory at unscheduled times to complete some exercises.

2482 Microbiology (3)

Prerequisite: Biology 1811 [majors must also take Biology 1821] and Chem 1111 [or Chem 1082 plus Chem 1091]. Study of microorganisms, their metabolism, genetics, and their interaction with other forms of life. Three hours of lecture per week.

2483 Microbiology Laboratory (2)

Prerequisite: Biology 2482 (may be taken concurrently). Experimental studies and procedures of microbiological techniques. Three and one-half hours of organized laboratory time per week. Students will need to return to the laboratory at unscheduled times to complete some exercises.

3622 Cell Biology (3)

Prerequisite: Biology 1811 [majors must also take Biology 1821], Chem 1111, 1121 and 2612 or equivalents. Examination of the basic biological processes of cells.

4602 Molecular Biology (3)

Prerequisite: Biology 2012 and 4712. A study of the principles of molecular biology, with emphasis on understanding the genetic regulation of DNA, RNA, and protein synthesis and function in the eukaryotic cells. Three hours of lecture per week. Students may not receive credit for both Biology 4602 and Biology 6602.

4612 Molecular Genetics of Bacteria (3)

Prerequisite: Biology 2482 and Biology 2012. A study of the molecular biology of gene replication, transfer, and expression in bacterial cells. Topics include DNA replication, transcription and translation, mutagenesis, DNA repair and recombination, gene transfer, and the regulation of genes and global expression systems. Three hours of lecture per week. Students may not received credit for both Biology 4612 and Biology 6612.

4614 Biotechnology Laboratory I (4)

Prerequisite: Biology 2012 or consent of instructor. An introduction to the fundamental concepts that underlie the field of biotechnology. Both the basic principles of molecular biology and hands-on experience with the techniques of the field will be addressed through lectures, discussion, and a series of laboratory exercises. Two hours of lecture and four hours of laboratory per week. Fulfills a laboratory requirement only; may not be used to fulfill the 4000-level or above lecture course requirement for the B.A. or B.S. degree in biology. Students may not receive credit for Biology 4614 and a comparable biotechnology course from another institution.

4615 Biotechnology Laboratory II (4)

Prerequisite: Biology 4614 and either Biology 4602 or Biology 4612, or consent of instructor. An in-depth look at theory and practice of biotechnology. Lectures and discussion will examine the underlying principles, and laboratory exercises will present hands-on experience with current techniques. One hour of lecture and six hours of laboratory per week. Fulfills a laboratory requirement only; may not be used to fulfill the 4000-5000 level lecture course requirement for the B.A. or B.S. degree in biology. Students may not receive credit for both biology 4615 and Biology 6615.

4622 Molecular Cell Biology (3)

Prerequisite: Biology 3622, Biology 4602, and Biology 4712 or consent of instructor. A study of the structural organization and processes of eukaryotic cells. Topics of discussion will include regulation of transcription, gene product processing and transport, organelle biogenesis and function, cytoskeletal structure and function, and cell interactions. Three hours of lecture per week. Students may not receive credit for both Biology 4622 and Biology 6622.

4632 Nucleic Acid Structure and Function (3)

Prerequisite: Biology 2012 and 4712 or equivalent or consent of instructor. Comprehensive view of structural properties of DNA and RNA that promote molecular interactions and biological function. Topics include physical properties of nucleic acids, formation and biological importance of higher order structures, RNA enzymatic activities, nucleic acid-protein interactions, and RNA metabolism. Three hours of lecture per week. Students may not receive credit for both Biology 4632 and 6632.

4712 Biochemistry (3)

[Same as Chemistry 4712]. Prerequisite: Chemistry 2612 and either Biology 1811 or Chem 2622. Examines the chemistry and function of cell constituents, and the interaction and conversions of intracellular substances. Students may not receive credit for both Biology 4712 and Chemistry 4712.

4713 Techniques in Biochemistry (2)

Prerequisite: Biology 4712 or Chemistry 4712 (may be taken concurrently). Laboratory activities introducing fundamental qualitative and quantitative biochemical techniques. Student evaluation will be based on laboratory participation, student laboratory reports, and written examinations. Three and one-half hours of organized laboratory time per week. Students my need to return to the laboratory at unscheduled times to complete some experiments.

4842 Immunobiology (3)

Prerequisite: Biology 4712 and Chem 2612. The fundamental principles and concepts of immunobiology and immunochemistry. Emphasis on the relation of immunological phenomena to biological phenomena and biological problems. Three hours of lecture per week.

5069 Topics in Cellular and Molecular Biology (1)

Prerequisite: Graduate standing. Presentation and discussion of student and faculty research projects and/or current research articles in molecular, cellular and developmental biology. May be repeated.

6602 Advanced Molecular Biology (3)

Prerequisite: Biology 2012 and 4712, or consent of instructor. A study of the principles of molecular biology, with emphasis on understanding the genetic regulation of DNA, RNA, and protein synthesis and function in eukaryotic cell. Three hours of lecture per week. Students will be required to give an oral presentation and/or write an extra paper on a topic relevant to the course. Students may not receive credit for both Biology 6602 and Biology 4602.

6612 Advanced Molecular Genetics of Bacteria (3)

Prerequisite: Biology 2012 and 2482. A study of the molecular biology of gene replication, transfer, and expression in bacterial cells. Topics include DNA replication, transcription and translation, mutagenesis, DNA repair and recombination, gene transfer, and the regulation of genes and global expression systems. Three hours of lecture per week. Students will be required to give an oral presentation and/or write an extra paper on a topic relevant to the course. Student may not receive credit for both Biology 6612 and Biology 4612.

6615 Advanced Biotechnology Laboratory II (4)

Prerequisite: Biology 4614 and either Biology 4602 or Biology 4612, or consent of instructor. An in-depth look at the theory and practice of biotechnology. Lectures and discussion will examine the underlying principles, and laboratory exercises will present hands-on experience with current techniques. One hour of lecture and six hours of laboratory per week. Students will be required to give an oral presentation and/or write an extra paper on a topic relevant to the course. Students may not receive credit for both Biology 6615 and Biology 4615.

6622 Advanced Molecular Cell Biology (3)

Prerequisite: Biology 4602, Biology 3622, and Biology 4712, or consent of instructor. A study of structural organization and processes of eukaryotic cells. Topics of discussion will include regulation of transcription, gene product processing and transport, organelle biogenesis and function, cytoskeletal structure and function, and cell interactions. Three hours of lecture per week. Students will be required to given an oral presentation and/or write an extra paper on a topic relevant to the course. Students may not receive credit for both Biology 6622 and Biology 4622.

6632 Advanced Nucleic Acid Structure and Function (3)

Prerequisite: Biology 2012 and 4712 or equivalent or consent of instructor. Comprehensive view of structural properties of DNA and RNA that promote molecular interactions and biological function. Topics include physical properties of mucleic acids, formation and biological importance of higher order structures, RNA enzymatic activities, nucleic acid-protein interaction, and RNA metabolism. Three hours of lecture and one hour of discussion per week. Students may not receive credit for both Biology 4632 and Biology 6632.

6642 Advanced Plant Molecular Biology and Genetic Engineering (3)

Prerequisite: Biology 4602 or 4612. Topics will include plant cell and developmental biology, DNA transfer into plants, using mutations to identify genes and their functions, regeneration of plants in tissue culture, signal transduction mechanisms, molecular biology of plant organelles, developmental engineering, metabolic engineering, plant microbe interactions, and engineered resistance to pathogen attack. Three hours of lecture and one hour of seminar per week. Students may not receive credit for both Biology 4642 and Biology 6642.

6652 Advanced Virology (3)

Prerequisite: Biology 2482 and 2012. An advanced comparative study of the structure, reproduction, and genetics of viruses. Three hours of lecture, one hour of discussion or seminar per week. Students may not receive credit for both Biology 4652 and 6652.

6699 Graduate Internship in Biotechnology (1-4)

Prerequisite: Graduate standing and enrollment in graduate Biotechnology Certificate Program. Six credit hours maximum (maximum of eight combined credit hours of Biology 6905 and internship). Internship will consist of period of observation, experimentation and on-the-job training in a biotechnology laboratory. The laboratory may be industrial or academic. Credit will be determined by the number of hours the student works each week and in consultation between the intern's supervisor and the instructor. Internship assignments will be commensurate with the education and experience of the student.

6889 Graduate Seminar (2)

Presentation and discussion of various research problems in biology. Graduate student exposure to the seminar process.

6905 Graduate Research in Biology (1-10)

Research in area selected by student in consultation with faculty members.

Chemistry Courses

1111 Introductory Chemistry I (5) [MS]

Prerequisite: Mathematics through college algebra and trigonometry may be taken concurrently. Presents an introduction to the fundamental laws and theories of chemistry. Laboratory experiments are designed to demonstrate some aspects of qualitative and quantitative analysis and to develop skills in laboratory procedures. Chemistry majors may not include both Chem 1082 and 1111, and both Chem 1011 and 1111 in ht e120 hours required for graduation. Three hours of lecture and one hour of discussion per week, one hour of laboratory-lecture and three hours of laboratory per week.

1121 Introductory Chemistry II (5) [MI, MS]

Prerequisite: Chem 1111 or advanced placement. Lecture and laboratory are a continuation of Chem 1111. Three hours of lecture and one hour of discussion per week; one hour laboratory-lecture and three hours of laboratory weekly.

2223 Quantitative Analysis (3) [C, MI, MS]

Prerequisite : Chem 1121. Principles and practice of elementary quantitative chemistry. The lecture treats descriptive statistics with emphasis on small samples; various types of competing equilibria pertaining to acid-base, complexometric and potentiometric titrations; and an introduction to spetrophotometric processes. The laboratory provides exercises in titrimetirc, gravimetric, and spectrophotometric techniques. Both portions of the course deal with the analytical chemistry of environmentally-significant problems. Two hours of lecture and four and one-half hours of laboratory weekly.

2612 Organic Chemistry I (3) [MS]

Prerequisite: Chem 1121. An introduction to the structure, properties, synthesis, and reactions of aliphatic and aromatic carbon compounds. Three hours of lecture per week.

2622 Organic Chemistry II (3) [MI, MS]

Prerequisite: Chem 2612. A systematic study of organic

reactions and their mechanisms; organic synthetic methods. Three hours of lecture per week.

2633 Organic Chemistry Laboratory (2) [C, MS]

Prerequisite: Chem 2612 or consent of instructor. An introduction to laboratory techniques and procedures of synthetic organic chemistry including analysis of organic compounds. One hour of lecture and four and one-half hours of laboratory per week.

3302 Physical Chemistry for the Life Sciences (3)

Prerequisites: Chemistry 2612 and Mathematics 1800 or Mathematics 1100, and Physics 1012. Principles and applications of physical chemistry appropriate to students pursuing degree programs in the life sciences. Topics will include thermodynamics, equilibria, kinetics, and spectroscopy. This course is intended for undergraduates seeking the B.S. degree in Biochemistry and Biotechnology and does not fulfill the physical chemistry requirement for other Chemistry B.A. and B.S. degree programs.

4712 Biochemistry (3)

[Same as Biology 4712] Prerequisite: Chem 2612 and either Biology 1811 or Chem 2622. The chemistry and function of cell constituents, and the interaction and conversions of intracellular substances. Three hours of lecture per week. Students may not receive credit for both Biology 4712 and Chemistry 4712. Biology 4712 may not be used to fulfill the 3000 or 4000 level lecture course requirement for the B.S. in Biology.

4722 Advanced Biochemistry (3)

Prerequisite: Chem 4712. Selected advanced topics in the chemistry of life processes. Three hours of lecture per week.

4733 Biochemistry Laboratory (2)

Prerequisite: Chem 4712 may be taken concurrently. Laboratory study of biochemical processes in cellular and subcellular systems with emphasis on the isolation and purification of proteins (enzymes) and the characterization of catalytic properties. One hour of lecture and three and one-half hour of laboratory per week.

4764 Interdisciplinary Topics in Biochemistry (3)

Prerequisite: Chem 4712; Chem 4722 strongly recommended. Includes advanced studies of enzyme mechanisms, the role of metal ions in enzymatic and non-enzymatic processes, and the application of computational chemistry to biological systems. Three hours of lecture per week.

4772 Physical Biochemistry (3)

Prerequisite: Chem 3312 or Chem/Bio 4712. Designed to acquaint students with concepts and methods in biophysical chemistry. Topics that will be discussed include protein and DNA structures, forces involved in protein folding and conformational stability, protein-DNA interactions, methods for characterization and separation of macromolecules, electron transfer, and biological spectroscopy. Three hours of lecture per week.

5794 Special Topics in Biochemistry (1-3)

Prerequisite: Consent of instructor. Selected topics in biochemistry. May be taken more than once for credit.

6787 Problem Seminar in Biochemistry (1)

Prerequisite: Consent of the biochemistry staff. Problems from the current literature, presentations and discussions by faculty, students and visiting scientists. Ph.D. students may take more than once for credit. Up to three credits may be applied to the M.S. degree program.

HIGHLIGHTS AND OBJECTIVES

Bachelor of Biotechnology and Genetic Engineering Program: The Bachelor in Biotechnology and Genetic Engineering (B.Sc.) degree program provides the students with sufficient theoretical knowledge and practical skills in different biotechnology disciplines required for the world of the 21st century.

Beside the academic foundation in essential science disciplines, the students will spend most of their time studying more specialized fields in biotechnology such as molecular biology, recombinant DNA, protein engineering, tissue culture, plant science, and environmental science and computing. Much of the learning will take place in the lab where students gain practical training in these fields using basic and sophisticated instrumentation and facilities.

All undergraduate students should spend 8 training weeks in one of the hospitals assigned by the Department. A faculty member follows the students, visit them while training and arrange lectures and seminars for them. Training Hospitals include: King Abdullah University Hospital, Princess Basma Hospital, King Hussein Medical Center, Jordan University Hospital, Prince Rashid Military Hospital, and Al-Basheer Hospital.

Students should complete 130 credit hours successfully to earn this degree. Students should pass the high school Exam/Scientific stream with 65% at least to be admitted to this program.

Program Goals

Students who are going to complete this program will acquire proficiency in the following areas:

1. Understanding of basic biotechnology theories.
2. Perform highly technical procedures fundamental

to biotechnology such as DNA extraction and characterization, cloning, polymerase chain reaction, blotting, ELISA, electrophoresis, tissue culture, transfection, protein isolation and purification.

3. Strength in one of the key areas (molecular biology, protein engineering, vaccine production, industrial microbiology and plant breeding) of biotechnology through conducting an independent research project in their final year.
4. Skills for effective writing, oral communication and managing laboratory activities.
5. Analyzing of protein and DNA sequencing using computer technology including the Internet and software designed for such purposes. In addition student will be able to conduct statistical analysis and graphical display.
6. Student will be aware of the ethics of Biotechnology.

Career Opportunities

Career prospects for biotechnology graduates is promising. With a strong practical training in laboratory skills, biotechnology graduates can be employed in different positions including:

1. Research assistant as part of a larger research team, conducting experiments in the laboratory or in the field.
2. Lab specialist in biotechnology, food, pharmaceutical, medical and agrotechnology industries
3. forensic and medical laboratory specialist. Examine DNA and other scientific evidence relating to criminal investigations and analyze DNA from clinical samples from patients for diagnosis of genetics diseases.

4. Environmental consultant by assess and monitor environmental contamination and help in developing and implementing bioremediation programs to clean up the environment.
5. Scientific journalist and media presenter: by writing articles and help in preparing broadcast programs to communicate the importance of scientific development to the public.
6. Sales representative / marketing professional, sell and provide advice about special scientific products to clients in research, help to solve specific problems related to equipment applications and experimental procedures.

MASTER OF SCIENCE IN BIOTECHNOLOGY

- Advanced Biotechnology Studies Program Chair named Howard Hughes Medical Institute Professor
- New Spring/Summer edition of BiotechnologyNEWS now available
- Hopkins Joins Successful International Conference on Emerging Biotechnology Applications, Global Regulatory Issues

Apply Now

Apply conveniently online or via mail.

Contact Us

Interested in applying? We can help answer your questions.

Request more information about this program or call us at 1-800-847-3330.

Biotechnology continues to expand rapidly with new discoveries and often life-saving products at a breathtaking pace. This industry, a merger of science and business, demands a multi-disciplinary workforce skilled in basic-research, product development, regulatory affairs and commercialization.

The Biotechnology industry harnesses advances in molecular biology, genomics, and proteomics to move discoveries and ideas out of the laboratory and into the product development pipeline. Students in the Master of Science in Biotechnology program learn to advance their careers by broadening their skillbase and understanding of this dynamic field.

Johns Hopkins offers students the ability to learn, advance and succeed in this exciting field with a variety of learning opportunities designed to meet the needs of working adults. So come inside—look around the website and then contact the Advanced Academic Programs office at the Baltimore Homewood Campus (410-516-6749) or Kirsty Gharavi at the Montgomery County Center (301-294-7162) to come in for an in-person visit or to consult with an advisor.

CERTIFICATE IN BIOTECHNOLOGY ENTERPRISE

- Advanced Biotechnology Studies Program Chair named Howard Hughes Medical Institute Professor
- New Spring/Summer edition of BiotechnologyNEWS now available
- Hopkins Joins Successful International Conference on Emerging Biotechnology Applications, Global Regulatory Issues

Apply Now

Apply conveniently online or via mail.

Contact Us

Interested in applying? We can help answer your questions.

Request more information about this program or call us at 1-800-847-3330.

Students who want a good understanding of the business aspects of biotechnology to advance in this rapidly changing field but are well versed in the scientific aspects can apply to the certificate program in Biotechnology Enterprise. Students can be admitted to the certificate program in Biotechnology Enterprise without prior graduate level work but a bachelor's degree in the life sciences is recommended.

For consideration, students submit the standard application form and official transcripts. A grade point average of 3.0 on a 4.0 scale is expected. Certificate requirements consist of six courses chosen from the list below. The four core Enterprise courses must be taken and two courses may come from the list of electives.

Students who successfully complete the certificate and subsequently decide to seek admission to the master's degree program in biotechnology will receive credit for five of the courses taken in the certificate. All time limit restrictions as noted in the admissions section of this catalog will apply.

NATIONAL CANCER INSTITUTE MOLECULAR TARGET AND DRUG DISCOVERY FELLOWSHIP

- NCI Fellows Abstract Titles for the Class of 2006: NCI Poster Titles
- Projects for Fellows Accepted in Fall 2004
- Advanced Biotechnology Studies Program Chair named Howard Hughes Medical Institute Professor

Apply Now

Apply conveniently online or via mail.

Contact Us

Interested in applying? We can help answer your questions.

Request more information about this program or call us at 1-800-847-3330.

Johns Hopkins University and the Center for Cancer Research/National Cancer Institute (CCR/NCI) have developed an innovative graduate program that prepares the next generation of scientists in drug discovery technologies. This new concentration in Molecular Targets and Drug Discovery Technologies combined with the CCR/NCI fellowship fully integrates the didactic training and hands-on laboratory experience required for graduates to contribute to the advancement of knowledge and research in the field of drug discovery.

Fellows earn an MS in Biotechnology with a concentration in Molecular Targets and Drug Discovery Technologies, participate in important cancer research, work in CCR/NCI laboratories, and receive paid tuition for up to two years and an annual stipend.

Up to five students per year will be selected for the two-year fellowship. Fellows receive the stipend only if they are accepted into the Master of Science in Biotechnology and the NCI Molecular Target and Drug Discovery Fellowship Program.

Applications for NCI Fellowship are due March 1, 2007 for the Fall 2007 class. Application to the Fellowship also requires an additional application to the MS in Biotechnology program.

BIOETHICS@

Biotechnology Ethics Courses at a Glance

Each course equals 2 staff development credits (for Iowa residents only) or 1 graduate credit. The online courses are for high school teachers, extension personnel, and others who educate youth and adult audiences about the ethical issues surrounding biotechnology. The courses require a minimum of 15 hours online, plus an additional 15 hours of reading, writing, and research. Course participants will be introduced to influential moral theories as tools to evaluate ethical arguments about biotechnology. All class work can be done on participants' home computers. The courses incorporate case studies for participants to experience and evaluate for use in their own classrooms. There will be extensive online discussion with other educators. By the end of each course, participants will be able to critically discuss and teach others about some moral issues surrounding particular biotechnologies. The ethics background learned also can be applied to other topics in the curriculum. Course instructed by Dr. Kristen Hessler, Iowa State University, and offered in cooperation with Area Education Agency 11, Johnston, Iowa, and Iowa State University, Ames. Each course is limited to 20 participants.

Tuition

Iowa State University graduate (one credit): $330

Two Staff develoment (two credits, Iowa residents only):

Course Content Preview

Courses are available for high school teachers, extension personnel, and others who educate youth and adult audiences about the issues surrounding biotechnology. Each module requires a minimum of 15 hours online, plus an additional 15 hours of reading, writing, and research.

• *(Course 1) July 10-28, 2006, Teaching Bioethics:*

This course will be of interest to any science or social studies teacher interested in teaching bioethics. Topics include: advantages of incorporating bioethics in science and social studies courses; appropriate objectives for bioethics units; a brief study of ethical theory; various approaches to bioethics pedagogy; how to relate bioethics issues to personal ethical issues familiar to students; how to help students identify ethical issues; how to address bioethics while respecting diversity of views in the classroom; and assessment for bioethics units.

• *Not Offered in Summer of 2006 (Course 2) Ethics and Animals:*

Prerequisite: Teaching Bioethics. This course will enable participants to recognize and distinguish different views about the moral status of animals; to articulate and defend their own ideas about the ethics of using animals for food, research, and education; and to incorporate ethical issues concerning animals in their courses. This course will benefit educators who discuss or use animals in their courses or outreach efforts, as well as social studies teachers interested in current controversies about society's uses of animals.

• *Not Offered in Summer of 2006 (Course 3) Ethics and Biotechnology:*

Prerequisite: Teaching Bioethics. Modern biotechnology is as controversial as it is promising. Teaching the associated ethical issues can help engage students to learn the relevant science concepts and to learn the skills necessary to contribute to ongoing social dialogue about science and society. Topics include an overview of ethical controversies about biotechnology and specific ethical issues in plant, animal, and human biotechnology.

Evaluation and Assignments

In addition to the activity report in the topics module, and the lesson plan or other educational activity, participants will be evaluated on their participation in course discussions.

Discussions

Discussions are asynchronous, meaning that participants post messages to discussion lists. This is a lot like using a bulletin board. The advantage of this is that participants do not have to find a time when everyone can log in all at once. However, because exchange of ideas is so important, participants will have to be working on the same topics at roughly the same time. It is not possible, therefore, for participants to work entirely at their own pace, for example by doing all coursework in the first few days of the course or by leaving all coursework until the end. Therefore, it is imperative that you be able to participate in the discussions on a regular basis during the course. If you have questions about whether the course will be flexible enough for your purposes, please contact the course instructor.

Important Dates

Login information will be mailed to participants the Friday before the course is scheduled to begin. Participants are encouraged to login as soon as they can thereafter in order to familiarize themselves with WebCT and begin to "meet" their colleagues.

Resources

eLearner: This site will help you decide if you're ready for an online course.

WebCT's computer and browser requirements.

The Stakeholder Concept

The stakeholder concept is one of the most attractive conceptual devices in business ethics. This concept has its origins in a theory of management, but has been seized upon by scholars in business ethics as a way of expressing the idea that businesses have obligations to a wide range of parties, beyond the stockholders to whom corporate heads were traditionally thought to be beholden. According to Freeman, the author responsible for popularizing the concept within the field of business ethics, stakeholders simply "are those groups who have a stake in or claim on the firm." Stakeholder theory has served as part of a reconciliation between forward-thinking business leaders, on one hand, and public interest groups and consumer advocates, on the other. It involves a reinterpretation of the social role of business, and a move away from the view, famously advocated by economist Milton Friedman, that the only proper function of corporate leaders is to increase the wealth of stockholders. The stakeholder concept, whatever its weaknesses, gives due credit to the fact that the motives of agents involved in business are relatively complex. The conflict between those varied motives has been a central theme in the business ethics literature. Freeman includes in his list of stakeholders suppliers, customers, employees, stockholders, and the firm's local community. This list, though typical of the lists given by stakeholder theorists, is not uncontroversial. Indeed, the stakeholder concept itself has its critics. Those critics charge that the stakeholder approach is incapable of guiding necessary improvements in corporate governance, that the multiple lines of accountability implied by acknowledging a multiplicity of stakeholders reduces efficiency, and that indeed the very idea of stakeholders as morally significant undermines the morally significant relationship between corporate executives and stockholders. Nonetheless, the stakeholder concept can be a useful one. In particular, the process known as "stakeholder analysis" can provide biotechnology companies with a lens through which to pay attention to the full range of interested parties. Stakeholder theory suggests that we should pay attention to the interests of any group or individual who is affected by, or who may affect, a decision or policy. A rich body of literature exists on the extent to which the interests of various stakeholders should be integrated into corporate decision-making, and ways in which that can be accomplished. I will not venture to take a position on just how long, or short, a list of relevant stakeholders would be. Nor will I stake a claim on the critical question of whether those stakeholders deemed relevant should merely have their wishes or needs accounted for, or whether they should instead be consulted directly. I will simply suggest that the concept of a stakeholder, and the understanding that a very wide range of individuals and groups have a stake in the workings of a given company, can serve as a focal point for discussing what obligations a company has, and to whom. Stakeholder analysis, in practice, is a relatively complex process, aimed at "identifying and understanding multiple (often competing) political, social, legal, economic, and moral claims of many constituencies." A simplified stakeholder analysis for a biotech firm, conducted in relation to its efforts at determining appropriate modes of benefit sharing, would proceed in approximately the following manner. The first, descriptive phase of the analysis would begin with listing the range of parties having a stake in the issue. These might include: employees, stockholders (if these exist), providers of biological resources (i.e., human research subjects), local communities, competing firms, firms in related fields (such as traditional pharmaceutical firms), government agencies (at home and abroad), and university-based researchers. All of these parties, and perhaps others, can be said to have a stake in the benefit sharing practices of a particular corporation. The next step in this descriptive process would be to assess the nature of each stakeholder's interest in this issue, perhaps categorizing such interests as ethical, political, economic, legal, and so on. The second, normative phase of the stakeholder analysis would include an assessment of the kinds and degrees of obligation that the corporation has to each stakeholder. With regard to some stakeholders, for example, the obligations will be written, contractual obligations. With regard to other stakeholders, they will be what are known as "fiduciary," or trust-based, obligations. Other stakeholder obligations might be grounded in theories of what good neighbours owe each other. In other cases, the conclusion may be that a particular stakeholder is owed no obligations at all. There is no straightforward, uncontroversial algorithm for determining the nature and extent of obligations to various stakeholders. The point here, however, is not to settle upon a canonical rank-ordering of the corporation's obligations to various stakeholders, but rather to accurately situate the corporation as enmeshed in a range of ethically significant relationships, and to begin the complex task of meeting the demands of those various relationships.

The stakeholder concept has, of course, already played a role in discussions related to biotech benefit sharing. For example, both the World Bank and the World Intellectual Property Organization have made use of the concept in discussion documents. But the use of the stakeholder concept in such discussions has occurred in relative isolation from the rich literature that exists concerning the plausibility and application of that concept. A more well-informed application of the concept is required.

Single Nucleotide Polymorphism (SNPs): Variations on a Theme

Wouldn't it be wonderful if you knew exactly what measures you could take to stave off, or even prevent, the onset of disease? Wouldn't it be a relief to know that you are not allergic to the drugs your doctor just prescribed? Wouldn't it be a comfort to know that the treatment regimen you are undergoing has a good chance of success because it was designed just for you? With the availability of millions of SNPs, biomedical researchers now believe that such exciting medical advances are not that far away. A SNP (pronounced "snip"), is a small genetic change, or variation, that can occur within a person's DNA sequence. The genetic code is specified by the four nucleotide "letters" A (adenine), C (cytosine), T (thymine), and G (guanine). SNP variation occurs when a single nucleotide, such as an A, replaces one of the other three nucleotide letters—C, G, or T.

An example of a SNP is the alteration of the DNA segment AAGGTTA to ATGGTTA, where the second "A" in the first snippet is replaced with a "T". On average, SNPs occur in the human population more than 1 percent of the time. Because only about 3 to 5 percent of a person's DNA sequence codes for the production of proteins, most SNPs are found outside of "coding sequences". SNPs found within a coding sequence are of particular interest to researchers because they are more likely to alter the biological function of a protein. Because of the recent advances in technology, coupled with the unique ability of these genetic variations to facilitate gene identification, there has been a recent flurry of SNP discovery and detection.

Finding single nucleotide changes in the human genome seems like a daunting prospect, but over the last 20 years, biomedical researchers have developed a number of techniques that make it possible to do just that. Each technique uses a different method to compare selected regions of a DNA sequence obtained from multiple individuals who share a common trait. In each test, the result shows a physical difference in the DNA samples only when a SNP is detected in one individual and not in the other. Many common diseases in humans are not caused by a genetic variation within a single gene but are influenced by complex interactions among multiple genes as well as environmental and lifestyle factors. Although both environmental and lifestyle factors add tremendously to the uncertainty of developing a disease, it is currently difficult to measure and evaluate their overall effect on a disease process. Therefore, we refer here mainly to a person's genetic predisposition, or the potential of an individual to develop a disease based on genes and hereditary factors. Genetic factors may also confer susceptibility or resistance to a disease and determine the severity or progression of disease. Because we do not yet know all of the factors involved in these intricate pathways, researchers have found it difficult to develop screening tests for most diseases and disorders. By studying stretches of DNA that have been found to harbor a SNP associated with a disease trait, researchers may begin to reveal relevant genes associated with a disease. Defining and understanding the role of genetic factors in disease will also allow researchers to better evaluate the role non-genetic factors—such as behavior, diet, lifestyle, and physical activity—have on disease. Because genetic factors also affect a person's response to drug therapy, DNA polymorphisms such as SNPs will be useful in helping researchers determine and understand why individuals differ in their abilities to absorb or clear certain drugs, as well as to determine why an individual may experience an adverse side effect to a particular drug. Therefore, the recent discovery of SNPs promises to revolutionize not only the process of disease detection but the practice of preventative and curative medicine.

SNPs and Disease Diagnosis

Each person's genetic material contains a unique SNP pattern that is made up of many different genetic variations. Researchers have found that most SNPs are not responsible for a disease state. Instead, they serve as biological markers for pinpointing a disease on the human genome map, because they are usually located near a gene found to be associated with a certain disease. Occasionally, a SNP may actually cause a disease and, therefore, can be used to search for and isolate the disease-causing gene. To create a genetic test that will screen for a disease in which the disease-causing gene has already been identified, scientists collect blood samples from a group of individuals affected by the disease and analyze their DNA for SNP patterns. Next, researchers compare these patterns to patterns obtained by analyzing the DNA from a group of individuals unaffected by the disease. This type of comparison, called an "association study", can detect differences between the SNP patterns of the two groups, thereby indicating which pattern is most likely associated with the disease-causing gene. Eventually, SNP profiles that are characteristic of a variety of diseases will be established. Then, it will only be a matter of time before physicians can screen individuals for susceptibility to a disease just by analyzing their DNA samples for specific SNP patterns.

SNPs and Drug Development

As mentioned earlier, SNPs may also be associated with the absorbance and clearance of therapeutic agents. Currently, there is no simple way to determine how a patient will respond to a particular medication. A treatment proven effective in one patient may be ineffective in others. Worse yet, some patients may experience an adverse immunologic reaction to a particular drug. Today, pharmaceutical companies are limited to developing agents to which the "average" patient will respond. As a result, many drugs that might benefit a small number of patients never make it to market. In the future, the most appropriate drug for an individual could be determined in advance of treatment by analyzing a patient's SNP profile. The ability to target a drug to those individuals most likely to benefit, referred to as "personalized medicine", would allow pharmaceutical companies to bring many more drugs to market and allow doctors to prescribe individualized therapies specific to a patient's needs.

NCBI's "Discovery Space" Facilitating SNP Research

Records in dbSNP are cross-annotated within other internal information resources such as PubMed, genome project sequences, GenBank records, the Entrez Gene database, and the dbSTS database of sequence tagged sites. Users may query dbSNP directly or start a search in any part of the NCBI discovery space to construct a set of dbSNP records that satisfy their search conditions. Records are also integrated with external information resources through hypertext URLs that dbSNP users can follow to explore the detailed information that is beyond the scope of dbSNP curation. Reproduced with permission from Sherry ST, Ward MH, Kholodov M, Baker J, Phan L, Smigielski EM, Sirotkin K."dbSNP: the NCBI database of genetic variation." *Nucleic Acids Research.* 2001; 29:308-311. To facilitate research efforts, NCBI's dbSNP is included in the Entrez retrieval system which provides integrated access to a number of software tools and databases that can aid in SNP analysis. For example, each SNP record in the database links to additional resources within NCBI's "Discovery Space".

Resources include: GenBank, NIH's sequence database; Entrez Gene, a focal point for genes and associated information; dbSTS, NCBI's resource containing sequence and mapping data on short genomic landmarks; human genome sequencing data; and PubMed, NCBI's literature search and retrieval system. SNP records also link to various external allied resources. Providing public access to a site for "one-stop SNP shopping" facilitates scientific research in a variety of fields, ranging from population genetics and evolutionary biology to large-scale disease and drug association studies. The long-term investment in such novel and exciting research promises not only to advance human biology but to revolutionize the practice of modern medicine.

SNPs and NCBI

Because SNPs occur frequently throughout the genome and tend to be relatively stable genetically, they serve as excellent biological markers. Biological markers are segments of DNA with an identifiable physical location that can be easily tracked and used for constructing a chromosome map that shows the positions of known genes, or other markers, relative to each other. These maps allow researchers to study and pinpoint traits resulting from the interaction of more than one gene. NCBI plays a major role in facilitating the identification and cataloging of SNPs through its creation and maintenance of the public SNP database (dbSNP). This powerful genetic tool may be accessed by the biomedical community worldwide and is intended to stimulate many areas of biological research, including the identification of the genetic components of disease.

REFERENCES

Bartley, D.M. 2005. Status of the world's fishery genetic resources.

Binns, A.N., Thomashow, M.F. 1998. Cell biology of *Agrobacterium* infection and transformation of plants. Annual Review of Microbiology 42:575-606.

Cardellino, R.A. 2005. Preparation of the first report on the State of the World's Animal Genetic Resources.

Daniell, H., Streatfield, S. J., Wycoff, K. 2001. Medical molecular farming: production of antibodies, biopharmaceuticals and edible vaccines in plants. Trends in Plant Science 6:219-226

De Vicente, M.C., Guzmán, F.A., Engels, J. and Ramanatha Rao, V. 2005. Genetic characterization and its use in decision-making for the conservation of crop germplasm.

Dhlamini, Z., Spillane, C., Moss, J.P., Ruane, J., Urquia, N. and Sonnino, A. 2005. Status of research and application of crop biotechnologies in developing countries: Preliminary assessment. FAO, Rome.

Diamond, J. 2002. Evolution, consequences and future of plant and animal domestication. Nature, 418, 700-707.

Drews, J.& Ryser, S. The role of innovation in drug development.

Druker, B. J. Imatinib alone and in combination for chronic myeloid leukemia. *Semin.Hematol.* 40, 50–8 (2003).

ERFP. 2003. Guidelines for the constitution of national cryopreservation programmes for farm animals. Publication No. 1 of the European Regional Focal Point on Animal Genetic Resources. Hiemstra, S.J. (ed.).

FAO. 1998a. The state of the world's plant genetic resources for food and agriculture.

FAO. 1998b. Secondary guidelines for development of national farm animal genetic resources management plans: Management of small populations at risk.

FAO. 1999. The global strategy for the management of farm animal genetic resources: Executive brief

FAO. 2000a. How appropriate are currently available biotechnologies for the forestry sector in developing countries?Background Document to Conference 2 of the FAO Biotechnology Forum (25 April to 30 June 2000):

FAO. 2000b. World watch list for domestic animal diversity.

FAO. 2001. Global forest resources assessment 2000.

FAO. 2004a. The state of world fisheries and aquaculture.

FAO. 2004b. Measurement of domestic animal diversity - a review of recent diversity studies. Information document prepared for the meeting of the Intergovernmental Technical Working Group on Animal Genetic Resources for Food and Agriculture, 31 March to 2 April 2004, Rome, Italy.

Ferreira, M.E. 2005. Molecular analysis of genebanks for sustainable conservation and increased useo f crop genetic resources.

Fischer, R., Twyman, R. M., Schillberg, S. 2003. Production of antibodies in plants and their use for global health. Vaccine 21:820-825.

Fuller, B. P. *et al.* Policy forum: Ethics — privacy in genetics research. *Science* 285, 1359–1361 (1999).

Gabriel, S. B. *et al.* The structure of haplotype blocks in the human genome. *Science* 296, 2225–2229 (2002).

Gardner, M. J. *et al.* Genome sequence of the human malaria parasite *Plasmodium falciparum*. *Nature* 419, 498–511 (2002).

Gavin, A. C. *et al.* Functional organization of the yeast proteome by systematic analysis of protein complexes. *Nature* 415, 141–147 (2002).

Goldstein, D.A., Thomas, J.A. 2004. Biopharmaceuticals derived from genetically modified plants. QJM: An International Journal of Medicine 97:705-716.

Golub, T.R.Genomic approaches to the pathogenesis of hematologic malignancy.*Curr.Opin.Hematol.* 8, 252–261 (2001).

Greer, D. and Harvey, B. 2004. Blue genes: Sharing and conserving the world's aquatic biodiversity. International Development Research Centre.

Guttmacher and Mark S.Guyer are at the National Human Genome Research Institute, National Institutes of Health, Bethesda, Maryland 20892, USA.

Guttmacher,A. E.& Collins, F. S. Genomic medicine — A primer.*N.Engl. J.Med.* 347, 1512–1520 (2002).

Hannon, G. J.RNA interference. *Nature* 418, 244–251 (2002).

Hanotte, O. and Jianlin, H. 2005. Genetic characterisation of livestock populations and its use in conservation decision-making.

Hiemstra, S.J., van der Lende, T. and Woelders, H. 2005. The potential of cryopreservation and reproductive technologies for animal genetic resources conservation strategies

Hilbert,D.Mathematical problems. *Bull.Am.Math. Soc.* 8, 437–479 (1902).

Hirschhorn, J.N., Lohmueller, K., Byrne, E. & Hirschhorn,K.A comprehensive review of genetic association studies. *Genet.*

Holt, R. A. *et al.* The genome sequence of the malaria mosquito *Anopheles gambiae*. *Science* 298, 129–149 (2002).

Horii, T., Nagao, Y., Tokunaga, T., and Imai, H. 20003. Serum-free culture of murine primordial germ cells and embryonic germ cells. Theriogen. 59, 1257-1264.

Horii, T., Nagao, Y., Tokunaga, T., and Imai, H. 20003. Serum-free culture of murine primordial germ cells and embryonic germ cells. Theriogen. 59, 1257-1264.

Jackson,D. A., Symons, R. H. & Berg, P. Biochemical method for inserting new genetic information into DNA of Simian Virus 40: circular SV40 DNA molecules containing lambda phage genes and the galactose operon of *Escherichia coli. Proc.Natl Acad. Sci.*

Kaczorowski, D.J., Patterson, E.S., Jastromb, W.E., Shamblott, M.J. 2002. Review Article. Glucose-responsive insulin-producing cells from stem cells. Diabetes Metab.

Kaczorowski, D.J., Patterson, E.S., Jastromb, W.E., Shamblott, M.J. 2002. Review Article. Glucose-responsive insulin-producing cells from stem cells. Diabetes Metab.

Kalka, C. et al. Transplantation of ex vivo expanded endothelial progenitor cells for therapeutic neovascularization. Proc. Natl. Acad. Sci. USA 2000; 97:3422-3427.

Kalkman EAJ, Bilgin YM, van Haren P, van Suylen R-J, Saxena PR, Schoemaker RG.Determinants of coronary reserve in rats subjected to coronary artery ligation or aortic banding. Cardiovasc Res 1996.

Kawano H, Do YS, Kawano Y, Starnes V, Barr M, Law RE, Hsueh WA. Angiotensin II has multiple profibrotic effects in human cardiac fibroblasts. Circulation 2000;101:1130-1137.

Kehat I, Kenyagin-Karsenti D, Snir M, Segev H, Amit M, Gepstein A, Livne E, Binah O, Itskovitz-Eldor J, Gepstein L. Human embryonic stem cells can differentiate into myocytes with structural and functional properties of cardiomyocytes. J Clin Invest. 2001; 108: 407-14.

Kellerman S, Moore JA, Zierhut W, Zimmer HG, Campbell J, Gerdes AM. Nuclear DNA content and nucleation patterns in rat cardiac myocytes from different models of cardiac hypertrophy. *J Mol Cell Cardiol* , 497-505 (1992).

Kennedy M, Firpo M, Choi K, Wall C, Robertson S, Kabrun N, Keller G. A common precursor for primitive erythropoiesis and definitive haematopoiesis.

Kerr, D.A., Lladó, J., Shamblott, M.J., Maragakis, N.J., Irani, D.N., Crawford, T.O, Krishnan, C., Dike, S., Gearhart, J.D., and

Rothstein, J.D. 2003. Human embryonic germ cell derivatives facilitate motor recovery of rats with diffuse motor neuron injury. J. Neurosci. 23, 5131-5140.

Kerr, D.A., Lladó, J., Shamblott, M.J., Maragakis, N.J., Irani, D.N., Crawford, T.O, Krishnan, C., Dike, S., Gearhart, J.D., and Rothstein, J.D. 2003. Human embryonic germ cell derivatives facilitate motor recovery of rats with diffuse motor neuron injury. J. Neurosci. 23, 5131-5140.

Kimura, T., Suzuki, A., Fujita, Y., Yomogida, K., Lomeli, H., Asada, N., Ikeuchi, M., Nagy, A., Mak, T.W., and Nakano, T. 2003. Development and Disease. Conditional loss of PTEN leads to testicular teratoma and enhances embryonic germ cell production. Development 130, 1691-1700.

Kimura, T., Suzuki, A., Fujita, Y., Yomogida, K., Lomeli, H., Asada, N., Ikeuchi, M., Nagy, A., Mak, T.W., and Nakano, T. 2003. Development and Disease. Conditional loss of PTEN leads to testicular teratoma and enhances embryonic germ cell production. Development 130, 1691-1700.

Klug MG, Soonpaa MH, Koh GY, Field LJ (1996) Genetically selected cardiomyocytes from differentiating embryonic stem cells form stable intracardiac grafts. J Clin Invest 98:216-224.

Kocher A, Schuster M, Szabolcs M, Itescu S. Cardiomyocyte regeneration after neovascularization of ischemic myocardium by human bone marrowderived angioblasts. Nature Medicine 2002, in press.

Kocher AA, Schuster MD, Szabolcs MJ, Takuma S, Burkhoff D, Wang J, Homma S, Edwards NM, Itescu S. Neovascularization of ischemic myocardium by human bone-marrow-derived angioblasts prevents cardiomyocyte apoptosis, reduces remodeling and improves cardiac function. Nature Med.

L Field: Future therapy for cardiovascular disease. In Proceedings of the NHLBI Workshop Cell Transplantation: Future Therapy for Cardiovascular Disease? Columbia, MD; 1998.

Labastie M-C, Cortes F, Romeo P-H, Dulac C, Peault B. Molecular identity of hematopoietic precursor cells emerging in the human embryo. Blood 1998;92:3624-3635.

Lancet 347:279.

Lanteri, S. and Barcaccia, G. 2005. Molecular markers based analysis for crop germplasm preservation.

Lenstra, J.A. and ECONOGENE Consortium. 2005. Evolutionary and demographic history of sheep and goats suggested by nuclear, mtDNA and Y-chromosome markers.

Levkau B, Koyama H, Raines EW, Clurman BE, Herren B, Orth K, Roberts JM, Ross R. Cleavage of p21cipl/wafl and p27 kipl mediates apoptosis in endothelial cells through activation of cdk2: role of a caspase cascade.

Li, Y., Jenkins, C.W., Nichols, M.A. and Xiong, Y. (1994) Cell cycle expression and p53 regulation of the cyclin-dependent kinase inhibitor p21. Oncogene, 9, 2261-2268 Steinman, R.A., Hoffman, B., Iro, A., Guillouf, C., Liebermann, D.A. and El- Houseini, M.E. (1994) Induction of p21 (WAF1/CIP1) during differentiation.

Liechty KW, MacKenzie TC, Shaaban AF, et al. Human mesenchymal stem cells engraft and demonstrate site-specific differentiation after in utero transplantation in sheep. Nat Med 2000, 6:1282-6.

Lin Y, Weisdorf DJ, Solovey A, Hebbel RP. Origins of circulating endothelial cells and endothelial outgrowth from blood. J Clin Invest 2000;105:71-77.

Lynch, H. T. & de la Chapelle, A. Genomic medicine: hereditary colorectal cancer.*N.Engl. J.Med.* 348, 919–932 (2003).

Lysaght, M.J., and Hazlehurst, A.L., "Private Sector Development of Stem Cell Technology and Therapeutic Cloning," *Tissue Engineering* 9(3): 555-561 (2003).

Ma, J.K-C., Drake, P.M.W., Christou, P. 2003. The production of recombinant pharmaceutical proteins in plants. Nature Reviews Genetics 4: 794-805.

MacLellan WR and Schneider MD. Genetic dissection of cardiac growth control pathways Annu. Rev. Physiol. 2000. 62:289-320.

Mahon NG, O'Roke C, Codd MB, et al. Hospital mortality of acute myocardial infarction in the thrombolytic era. Heart 1999;81:478-82.

Maier S, Tertilt C, Chambron N, Gerauer K, Huser N, Heidecke C-D, Pfeffer K (2001) Inhibition of natural killer cells results in acceptance of cardiac allografts in CD28-/- mice. Nature Med 7:557-562.

Makino S, Fukuda K, Miyoshi S, et al. Cardiomyocytes can be

generated from marrow stromal cells in vitro. J Clin Invest 1999, 103:697-705.

Maxam, A. M. & Gilbert,W. A new method for sequencing DNA.

McCarthy PM, Rose EA, Macmanus Q, et al. Clinical experience with the Novacor ventricular assist system. J Thorac Cardiovasc Surg. 1991; 102:578-587.

McEwan PE, Gray GA, Sherry L, Webb DJ, Kenyon CJ. Differential effects of angiotensin II on cardiac cell proliferation and intramyocardial perivascular fibrosis in vivo. Circulation 1998;98:2765-2773.

Mendel, G.Versuche über Pflanzen-Hybriden.*Verhandlungen des naturforschenden Vereines, Abhandlungen, Brünn* 4, 3–47 (1866).

Migeon, B.R., Chowdry, A.K., Dunston, J.A., and McIntosh, I. 2001. Identification of *TSIX*, encoding an RNA antisense to human *XIST*, reveals differences from its murine counterpart: Implications for X inactivation. Am. J. Human Genet. 69, 951-960.

Migeon, B.R., Chowdry, A.K., Dunston, J.A., and McIntosh, I. 2001. Identification of *TSIX*, encoding an RNA antisense to human *XIST*, reveals differences from its murine counterpart: Implications for X inactivation. Am. J. Human Genet. 69, 951-960.

Milki, A.A. et al., "Incidence of monozygotic twinning with blastocyst transfer compared to cleavage-stage transfer" *Fertility and Sterility* 79(3): 503-506 (2003).

Miller, P. S. Is there a pink slip in my genes? *J.Health Care Law Policy* 3, 225–265 (2000).

National Center for Health Statistics, "Births: Final Data for 2001." National Vital Statistics Reports 51(2) (2002), available at http://www.cdc.gov/nchs/data/nvsr/ nvsr51/nvsr51_02.pdf.

National Institutes of Health Revitalization Act of 1993, Pub. L. No. 103-43, § 121(c), 107 Stat. 122 (1993), repealing 45 C.F.R. § 46.204(d).

National Institutes of Health, *Report of the Human Embryo Research Panel*, Bethesda, MD: NIH (1994).

National Research Council.*Mapping and Sequencing the Human Genome* (National Academy Press,Washington DC, 1988).

Nelissen-Vrancken H, Debets J, Snoeckx L, Daemen M, Smits J.

Time-related normalization of maximal coronary flow in isolated perfused hearts of rats with myocardial infarction. Circulation 1996;93:349-355.

Nester, E.W., Gordon, M.D., Amasino, R.M., Yanofsky, M.F. 1984. Crown gall: a molecular and physiological analysis. Annual Review of Plant Pathology 35:387-413.

Nidorf SM, Siu SC, Galambos G, Weyman AE, Picard MH. Benefit of late coronary reperfusion on ventricular morphology and function after myocardial infarction. J Am Coll Cardiol 1992;20:307-313.

Nirenberg, M.W. The genetic code: II. *Sci.Am.* 208, 80–94 (1963).

Norwitz, E. R., et al., "Implantation and the survival of early pregnancy," *New England Journal of Medicine* 345(19): 1400-1408 (2001)

Ogawa M, Kizumoto M, Nishikawa S, Fujimoto T, Kodama H, Nishikawa SI. Blood 1999;93:1168-1177.

Oncogene, 9, 3389-3396.

Onyango, P., Jiang, S., Uejima, H., Shamblott, M.J., Gearhart, J.D., Cui, H., and Feinberg, A.P. 2002. Monoallelic expression and methylation of imprinted genes in human and mouse embryonic germ cell lineages. Proc. Natl. Acad. Sci. USA 99, 10599-10604.

Onyango, P., Jiang, S., Uejima, H., Shamblott, M.J., Gearhart, J.D., Cui, H., and Feinberg, A.P. 2002. Monoallelic expression and methylation of imprinted genes in human and mouse embryonic germ cell lineages. Proc. Natl. Acad. Sci. USA 99, 10599-10604.

Orlic D, Kajstura J, Chimenti S, et al. Bone marrow cells regenerate infarcted myocardium. Nature 2001, 410:701-705.

Oz MC, Argenziano M, Catanese KA, et al. Bridge experience with long-term implantable left ventricular assist devices. Are they an alternative to transplantation? Circulation. 1997; 95:1844-1852.

Panis, B. and Lambardi, M. 2005. Status of cryopreservation technologies in plants (crops and forest trees).

Papa, R. 2005. Gene flow and introgression between domesticated crops and their wild relatives.

Park, T.S., Hong, Y.H., Kwon, S.C., Lim, J.M., and Han, J.Y. 2003. Birth of germline chimeras by transfer of chicken embryonic germ (EG) cells into recipient embryos.

Park, T.S., Hong, Y.H., Kwon, S.C., Lim, J.M., and Han, J.Y. 2003. Birth of germline chimeras by transfer of chicken embryonic germ (EG) cells into recipient embryos.

Peterson, R.K.D., Arntzen, C.J. 2004. On risk and plant-based biopharmaceuticals. Trends in Biotechnology 22:64-66.

Pfeffer JM, Pfeffer MA, Fletcher PJ, Braunwald E. Progressive ventricular remodeling in rat with myocardial infarction. Am J Physiol 1991; 260:H1406-14.

Pfeffer MA, Braunwald E, Moye LA, et al. Effect of captopril on mortality and morbidity in patients with left ventricular dysfunction after myocardial infarction. Results of the survival and ventricular enlargement trial. The SAVE investigators. N Engl J Med 1992;327:669-677.

Pitt B, Segal R, Martinez FA, et al. Randomised trial of losartan versus captopril in patients over 65 with heart failure (Evaluation of Losartan in the Elderly Study, ELITE). Lancet 1997;349:747-752.

Pittenger MF, Mackay AM, Beck SC, et al. Multilineage potential of adult human mesenchymal stem cells. Science 1999, 284:143-147.

Pouzet B, Ghostine S, Vilquin J-T, Garcin I, Scorsin M, Hagege AA, Duboc D, Schwartz K, Menasche P (2001) Is skeletal myoblast transplantation clinically relevant in the era of angiotensin-converting enzyme inhibitors? Circulation 104:I223-228.

Primmer, C. 2005. Genetic characterisation of populations and its use in conservation decision making in fish.

Purdue Agricultural Biotechnology. 2004. Moving Genes into Plants. http://www.agriculture.purdue.edu/agbiotech/genetransfer.html

Rafii S, Shapiro F, Rimarachin J, Nachman R, Ferris B, Weksler B, Moore AS, Asch AS. Isolation and characterization of human bone marrow microvascular endothelial cells: hematopoietic progenitor cell adhesion.

Rao, N.K. 2004. Plant genetic resources: Advancing conservation and use through biotechnology. African Journal of Biotechnology, 3, 136-145.

Ravichandran LV and Puvanakrishnan R. In vivo labeling studies on the biosynthesis and degradation of collagen in experimental myocardial myocardial infarction. Biochem Intl 1991;24:405-414.

Reich,D. E. & Lander, E. S. On the allelic spectrum of human disease. *Trends Genet.* 17, 502–510 (2001).

Rogers, K.K. 2003. The Potential of Plant-Made Pharmaceuticals. http://www.plantpharma.org/ials/index.php?id=1

Rothenberg, K. *et al.* Genetic information and the workplace: Legislative approaches and policy challenges. *Science* 275, 1755–1757 (1997).

Ruane, J. 2000. A framework for prioritizing domestic animal breeds for conservation purposes at the national level: a Norwegian case study. Conservation Biology, 14, 1385-1393.

Sachidanandam, R. *ct al.* A map of human genome sequence variation containing 1.42 million single nucleotide polymorphisms. *Nature* 409, 928–933 (2001).

Sala, F., Rigano, M. M., Barbante, A., Basso, B., Walmsley, A.M., Castiglione, S. 2003. Vaccine antigen production in transgenic plants: strategies, gene constructs and perspectives. Vaccine 21:803-808.

Sanger, F.& Coulson, A. R. A rapid method for determining sequences in DNA by primed synthesis with DNA polymerase.

Sapienza, C. 2002. Commentary. Imprinted gene expression, transplantation medicine, and the "other" human embryonic stem cell. Proc. Natl. Acad. Sci. USA 99, 10243-10245.

Sapienza, C. 2002. Commentary. Imprinted gene expression, transplantation medicine, and the "other" human embryonic stem cell. Proc. Natl. Acad. Sci. USA 99, 10243-10245.

Selkoe,D. J.Alzheimer's disease: genes, proteins, and therapy. *Physiol. Rev.* 81, 741–66 (2001).

Shamblott M.J., et al., "Derivation of pluripotent stem cells from cultured human primordial germ cells" *Proceedings of the National Academy of Sciences.* 95(23): 13726-13731 (1998).

Shi Q, Rafii S, Wu MH-D, et al. Evidence for circulating bone marrow-derived endothelial cells. Blood 1998;92:362-367.

Sidow, A. Sequence first.Ask questions later. *Cell* 111, 13–16 (2002).

Sigaud, P. 2005. Efforts towards assessing the global status of forest genetic resources.

Simianer, H. 2005. Use of molecular markers and other information for sampling germplasm to create an animal genebank.

Smith, L. M. *et al.* Fluorescence detection in automated DNAsequence analysis. *Nature* 321, 674–679 (1986).

Smythe GM, Grounds MD (2000) Exposure to tissue culture conditions can adversely affect myoblast behaviour in vivo in whole muscle grafts: implications for myoblast transfer therapy. Cell Transplant 9:379-393.

Smythe GM, Hodgetts SI, Grounds MD (2001) Problems and solutions in myoblast transfer therapy. J Cell Mol Med 5:33-47.

Soonpaa MH, Field LJ. Assessment of cardiomyocyte DNA synthesis in normal and injured adult mouse hearts. *Am J Physiol* , H220-6 (1997).

Stockwell, B. R. Chemical genetics: Ligand-based discovery of gene function. *Nature Rev. Genet.* 1, 116–125 (2000).

Streatfield, S.J., Lane, J.R., Brooks, C.A., Barker, D.K., Poage, M.L., Mayor, J.M., Lanphear, B.J., Drees, C.F., Jilka, J.M., Hood, E.E., Howard, J.A. 2003. Corn as a production system for human and animal vaccines. Vaccine 21:812-815.

Suzuki K, Smolenski R.T., Jayakumar J, Murtuza B., Brand N.J., Yacoub MH (2000) Heat shock treatment enhances graft cell survival in skeletal myoblast transplantation to the heart. Circulation 102:III216-221.

Takahashi, T. et al. Ischemia- and cytokine-induced mobilization of bone marrowderived endothelial progenitor cells for neovascularization. Nat. Med. 1999; 5:434-438.

Thomas, B.R. Van Deynze, A., Bradford, K.J. 2002. Production of Therapeutic Proteins in Plants. Agricultural Biotechnology in California Series 8078. University of California-Davis. 12pp.

Walmsley, A.M., Arntzen, C.J. 2000. Plants for delivery of edible vaccines. Current Opinion in Biotechnology 11:126-129.

Appendix

EDUCATIONAL AND INDUSTRIAL SOURCES OF BIOTECHNOLOGY

Agricultural Biotechnology

The United States Department of Agriculture (USDA) is one of three Federal Agencies, along with the Environmental Protection Agency (EPA) and the U.S. Food and Drug Administration (FDA), primarily responsible for regulating biotechnology in the United States. Products are regulated according to their intended use, with some products being regulated under more than one agency.

h*ttp://www.usda.gov/agencies/biotech/*

Aurora Biosciences Corporation

Aurora Biosciences Corporation designs, develops and commercializes proprietary drug discovery systems and technologies to accelerate and enhance discovery of new medicines by the pharmaceutical and biopharmaceutical industries. The Company is developing a proprietary biology and high technology platform comprised of a portfolio of proprietary functional genomics technologies, and a highly automated ultra-high throughput screening system, known as the Company's UHTSS Platform.

http://www.aurorabio.com/

BioABACUS

Biotechnology Abreviation and Acronym Uncovering Service (BioABACUS) is a searchable database of abbreviations and acronyms in Biotechnology that contains terms in such categories as: Biochemistry, Cell Biology, Computers and Internet, Diseases, Grants, Journals, Laboratories, Medicine, Molecular Biology/Genetics, Neuroscience, Other Organizations, Professional Societies and US Government. The data provided include the name of the term, its meaning, the category in which it is mostly used and, when available, a link in which additional information for the term can be found.

http://www.nmsu.edu/~molbio/bioABACUShome.htm

BioInformatics Services (BIS)

BIS is a biotechnology firm specializing in applying the power of computational biology to build solutions to complex biological and data analysis problems faced by scientists in public and private research institutes, biotechnology companies, and universities.

http://www.bioinformaticsservices.com/

Biology and Biotechnology Research Program at the Lawrence Livermore National Laboratory

The Biology and Biotechnology Research Program is focused on three strategic thrust areas: genomics, disease susceptibility and prevention, bioengineering research. These thrust areas provide a set of core compentencies and a foundation for our laboratory-wide strategic plan in the biosciences.

http://www-bio.llnl.gov/bbrp/bbrp.homepage.html

BioMetics, Inc.

BioMetics, Inc. provides engineering, design, validation, cGMP consulting and project management services for biopharmaceutical and environmental technology companies. The company delivers the highest quality of services, adding value to each project through innovative approaches to design and cost containment.

http://www.biometicsma.com/

Bio Online

Bio Online is a comprehensive site for biotechnology-related information and services. The site contains current news, featured articles, events, industry reports and analyses, press releases, product information, and much more. Bio Online exists to create a distributed on-line community of scientists, professionals, businesses and organizations supporting the life sciences for the purpose of facilitating communications and disseminating information within that community. Opening channels of information can only benefit those who strive to contribute to these endeavors.

http://www.bio.com/os/start/home.html

BioSpace.com

A global hubsite for the life sciences, including current news, industry events, resources, search tools, and much more, to help foster a worldwide community of shared interests within the biotech industry.

http://www.biospace.com/

BioTech's Biotechnology Dictionary

A searchable online dictionary of biology, chemistry and biotechnology terms. This free resource is still very much under construction. Currently, most of the 8300+ terms deal with biochemistry, biotechnology, botany, cell biology and genetics. There are also some terms relating to ecology, limnology, pharmacology, toxicology and medicine. Don't expect to find common or exotic animals here; there are far too many species to cover even a fraction in a resource like this. However, medically- and biotechnologically-relevant organisms such as bacteria, worms, fungi, and some plants are included. Looking for a chemical acronym? Try searching BioTech's Chemical Acronyms Database.

http://biotech.icmb.utexas.edu/search/dict-search.html

Bio-Technical Resources

Bio-Technical Resources provides innovative solutions for the development of biotechnology products and processes through contract services in research and development, bioprocess scale-up, pilot scale manufacturing, technology and economic assessments.

http://www.biotechresources.com/

Biotechnology and Biological Sciences Research Council (BBSRC)

The BBSRC is the leading funding agency for academic research and training in the biosciences at universities and institutes throughout the UK. The research that funded aims to improve the understanding of living organisms and biological systems in seven major areas: agri-food, animal sciences, biochemistry and cell biology, biomolecular sciences, engineering and biological systems, genes and developmental biology, and plant and microbial sciences.

http://www.bbsrc.ac.uk/

Biotechnology Group of Danish Institute of Agricultural Sciences

Biotechnology Group of Danish Institute of Agricultural Sciences, DIAS, is a small research team engaged in plant breeding related research. Promoting the use of transgenic crop plants in agriculture is at the core of all research activities of the group. All projects comprise a mix of applied and fundamental science, and the Group is thus an important link between applied oriented DIAS and the Danish universities.

http://www.dias.kvl.dk/

Biotechnology Program at UC Davis

UC Davis has long enjoyed a reputation as one of the world's leading institutions for research in biological sciences. UC Davis houses leading departments in each of the major areas where biotechnology has a significant impact: agriculture covering plant, animal, microbial science, entomology and sustainability; medicine and veterinary medicine including therapeutics, diagnostics and pathology; food science; bioprocess technology and environmental science. Research in molecular and cellular biology and biochemistry has led to the development of new research approaches and the new tools of biotechnology. UC Davis views these recent research results as part of a progressive cycle, with technology contributing to the advancement of research goals and biological research contributing, in turn, to the advancement of technology.

http://www.biotech.ucdavis.edu/

BioTime, Inc.

BioTime, Inc. is engaged in the research and development of aqueous-based synthetic solutions that may be used as plasma expanders, organ preservation solutions, or solutions to replace blood volume during low temperature surgery.

http://www.biotimeinc.com

BioValley

The principle goal of BioValley is the promotion of greater cooperation between companies involved in the biotechnological and biomedical sectors and the internationally respected scientific institutions associated with universities in the BioValley area, most of which have already formed close links between themselves. The intention is to address not only the many pharmaceutical concerns already present in the BioValley area, but also to involve the region's numerous smaller enterprises and suppliers. This calls for the creation of a network strongly committed to technology transfer.

http://www.biovalley.com/

Boston Biomedica, Inc.

Boston Biomedica, Inc. is an infectious diseases management company providing products and services for the detection and treatment of infectious diseases (ID) such as AIDS, Lyme Disease, and Viral Hepatitis. Boston Biomedica is also pursuing research and development programs in other areas, including Pressure Cycling Technology (PCT), a new technology which may allow more precise control chemical of reactions through the use of pressure.

http://www.bbii.com/

Cell Genesys Inc.

Cell Genesys is engaged primarily in the development and commercialization of gene therapies to treat major, life-threatening diseases, including cancer and AIDS. Cell Genesys currently has two gene therapy programs, the clinical programs and the pre-clinical programs. The clinical programs include GVAX(tm) cancer vaccines in Phase I/II studies to treat specific types of cancer, such as lung and prostate cancers, and T cell gene therapy for AIDS, which is undergoing Phase II testing.

http://www.cellgenesys.com/

Cell Trends Inc.

Cell Trends Inc. is a biotechnology firm that offers technical and research project work to scientists of virtually every level of sophistication: industry, government agencies, medical schools, colleges, and universities. Cell Trends offers a wide range of cell based laboratory services, utilizing a diverse selection of mammalian and insect cell lines. Features include baculovirus gene expression, DNA recombinant technology, soluble and membrane bound protein production (using hollow fiber bioreactors, roller bottles, or spinner flasks), gene cloning, transformant selection, electrophoresis, PCR, ELISA, and selection of immortalized cell lines.

http://www.celltrends.com/

Center for Advanced Biotechnology and Medicine (CABM)

CABM was established in response to a recommendation in the report of the Governor's Commission on Science and Technology that New Jersey should create and maintain a world-class center in biotechnology. Launched in 1985, CABM now encompasses the Core Facility, conducting basic research in the areas of structural biology, molecular genetics, and cell and developmental biology; and educational Programs, including lectures, seminars, symposia, short courses, fellowships, and visiting scientist programs.

http://www.cabm.rutgers.edu/

Center for Advanced Research in Biotechnology (CARB)

CARB, located at the University of Maryland, was designed to meet the specific needs of the biotechnology community in the field of protein structure, function, and design. Scientists from the Biotechnology Division of the Chemical Science and Technology Laboratory of NIST and the University of Maryland have come together to conduct research programs in structural and physical biochemistry and molecular biology. CARB, with sophisticated state-of-the-art instrumentation, provides an exceptional environment for training graduate students and postdoctoral fellows in fields critical to the future of biotechnology. CARB offers a Visiting Scientists Program that allows researchers from industrial, academic, and governmental laboratories to work at the facility for extended periods.

http://indigo15.carb.nist.gov/

Center for Biotechnology at Oak Ridge National Laboratory

Biotechnology is an integrating science that emphasizes the development and application of biological approaches, or the use of engineering and computational capabilities, to meet national needs in health and environmental protection, biomedical applications, environmental remediation, pollution abatement, and energy production. ORNL is uniquely positioned in this emerging field owing to its diversity of interactive disciplinary foundations (e.g., biology, biochemistry, molecular biology, computational sciences, chemistry, ecology, bioengineering, and biophysics) and its core competency in biological and environmental science and technology. The ORNL

Center for Biotechnology provides a coordinated focus for this biotechnology program, integrating activities across 14 ORNL divisions and forming partnerships with other institutions, both public and private, throughout the world.

http://www.ornl.gov/cbt/

ChemGen

ChemGen is a privately held biotechnology company specializing in developing products through the use of microbial fermentation technology. The company's focus has been on the production of enzymes for patented applications and on chemicals, including pharmaceutical intermediates, using unique production processes.

http://www.chemgen.com/

ChemGenes Corporation

ChemGenes Corporation, a biotechnology company engaged in the field of DNA/RNA synthesis. ChemGenes holds a number of patents in the field of standard, modified, as well as backbone modified DNA/RNA synthesis products.

http://www.chemgenes.com/

Cleveland Biotech Ltd.

Cleveland Biotech grows and acclimatises naturally occurring bacteria to create products for the degradation of organic wastes, including fats, oil and greases, cellulose, petroleum hydrocarbons, and chlorinated organics. Products are also available for the degradation of inorganic pollutants such as ammonia and nitrite.

http://www.clevebio.com/

Corixa Corporation

Corixa Corporation is a research and development-based biotechnology company committed to treating and preventing autoimmune diseases, cancer and infectious diseases by understanding and directing the immune system. The Company is focused on immunotherapeutic products and has a broad technology platform enabling both fully integrated vaccine design and the use of its separate proprietary vaccine components (antigens, adjuvants or antigen delivery technology) on a standalone, "Powered by Corixa" basis.

http://www.corixa.com/

Council for Biotechnology Information

Leading biotechnology companies have founded Council for Biotechnology Information to create a public dialogue and to share information about biotechnology that is based on objective scientific research, indcpendent expert opinion, and peer-reviewed published reports. The web site is also available in the French language.

http://www.whybiotech.com/

Covance Inc.

Covance's biotechnology services include capabilities in cGMP manufacture of monoclonal antibodies, growth factors, cytokines, and other recombinant products for clinical trials and commercial use. The company offers a comprehensive range of experience developing biopharmaceutical production processes leading to optimal expression, purification, formulation, and testing of recombinant products. Equally important, Covance can effectively transfer established manufacturing processes into large-scale cGMP manufacturing areas and provide a reliable supply of the highest quality recombinant product.

http://www.covance.com/biotech/

CSL Limited

CSL Limited is an Australian public company specialising in the development, manufacture and marketing of biologically based health care products which benefit the community. Products include human and veterinary pharmaceuticals (notably vaccines) and products derived from human plasma, diagnostics, and cell culture reagents. CSL is firmly committed to research and development, quality assurance and the development of international product and marketing alliances.

http://www.csl.com.au/

CuraGen Corporation

CuraGen Corporation is an integrated, genomics-based drug discovery and development company that is revolutionizing the Life Sciences industry by merging biology with information technology. CuraGen has created an integrated suite of genomics technologies to better understand the role of genes in improving human health, animal health and agricultural vitality. This technology is being applied to augment the discovery and development efforts of the Company's collaborators, while simultaneously advancing the development of CuraGen's internal pipeline of protein therapeutics and antibody drugs for the treatment of complex diseases.

http://www.curagen.com/

Dyadic International Inc.

Dyadic is a leader in the development and production of industrial enzymes and speciality chemicals, holding US and International patents. Currently the company's *non-GMO* products are used in the textile, food, starch and animal feed industries. An area targeted for future development is the pulp and paper industry.

http://www.dyadic-group.com/

Enviro Research Pty Ltd

Enviro Research has been engaged in microalgal culture research for a number of years. The company has developed the world's first and only closed photobioreactor which can be scaled up to any volume. The Envirocell offers all the control, hygiene, efficiency and reliability of laboratory based culture systems at industrial volumes. The Envirocell is currently the best microalgal culture system in the world for the aquaculture industry, fine chemicals/pharmaceuticals companies, health food/nutraceuticals industry, and wastewater nutrients/toxins removal industry.

http://www.enviro-research.com.au/

Epimmune Inc.

Epimmune develops therapeutic products for treatment of immunological diseases. The company's core business is the development of vaccines to treat and prevent infectious diseases and cancer. Epimmune's business also includes a therapeutic anti-inflammatory program based on cell adhesion inhibitors and its Glytec business unit, encompassing proprietary manufacturing technology for the enzymatic synthesis of bioactive complex carbohydrates.

http://www.epimmune.com/start.html

Exiqon A/S

Exiqon is a Danish biotech company focused on the development of enabling technology platforms and novel DNA- and immuno-diagnostic products against common genetic diseases, in particular, cardio-vascular, diabetic and other chronic and degenerative diseases. In other fields Exiqon is a potential technology partner for licensees. Exiqon's IPR includes LNA (Locked Nucleic Acids) and a universal method for the covalent coupling of diagnostic molecules to solid supports.

http://www.exiqon.com/

Federal Bio-Technology Transfer Directory

The Federal Bio-Technology Transfer Directory is an integrated information resource designed for users in the biotechnology, pharmaceutical and healthcare industries. Besides being essential for access to technology transfer opportunities, the database is a unique source for competitive intelligence concerning federal (and corporate) research, development and commercialization activities.

http://www.bioinfo.com/fbdinfo.html

Gemini Biotech

Gemini Biotech provides prompt, high quality, and cost-effective support for research and drug discovery. The company's technical staff and consultants are experts in synthetic chemistry and molecular biology. Gemini Biotech specializes in custom DNA probes/primers, modified oligonucleotides and DNA-RNA hybrids, full-length genes, gene sequencing, custom peptides and modifications, peptide conjugates, and peptide antibodies.

http://www.geminibio.com/

Genetics Forum

The Genetics Forum is an independent organisation in the UK concerned with the use of new genetic technologies and their public policy implications. It was founded by a group of scientists, lawyers and advocates from the animal welfare, environmental and consumer movements concerned about the long-term impact of rapid developments in the genetic sciences.

http://www.geneticsforum.org.uk/

Genetics Institute

Genetics Institute is a leading biopharmaceutical company engaged in the discovery, development and commercialization of human pharmaceuticals through recombinant DNA and other technologies. The company has a diversified portfolio of licensed and proprietary pharmaceutical products at various stages of development, including treatments for anemia, hemophilia, cancer, bone damage, inflammatory conditions and immune system disorders.

http://www.genetics.com/genetics/genetics/index.htm

Genzyme Transgenics Corporation

Genzyme Transgenics Corporation, a biotechnology company, is a leader in the production of therapeutic proteins in the milk of transgenic animals. Transgenic animals carry specific genetic material that allows them to produce human proteins in their milk. Unique molecular biology and production expertise has enabled Genzyme Transgenics to produce more than 60 therapeutic proteins, including monoclonal antibodies, plasma proteins and certain hard-to-express proteins in the milk of mice, rabbits, goats and cows. The company has developed substantial expertise and increasing focus on monoclonal antibodies for therapeutic use in chronic diseases such as rheumatoid arthritis, other autoimmune conditions and cancer.

http://www.transgenics.com/

Geron Corporation

Geron is a biopharmaceutical company that is focused on discovering, developing and commercializing therapeutic and diagnostic products for applications in oncology, drug discovery and regenerative medicine. Geron's product development programs are based upon three patented, independent and synergistic technologies: telomerase, human pluripotent stem cells, and nuclear transfer.

http://www.geron.com/

Greenpeace: Genetic Engineering

This site, which is operated by Greenpeace, advocates no genetic manipulation of nature. It focuses on (1) promoting natural foods and the segregation of genetically engineered foods with proper labels, (2) preventing genetically modified organisms from being released into the environment, (3) preventing any living organisms or their genes being subject to private property claims (i.e., no patents), and (4) promoting genetic engineering issues being addressed by international bodies, with international regulations based on the precautionary principle. Greenpeace USA, a part of the international organization dedicated to protecting the health of planet Earth, offers on-line media kits providing background information on various issues, including Greenpeace's concern that the FDA's safety standards regarding genetically engineered foods are insufficient and may be putting human health at risk. This media kit links to Greenpeace's report titled *FDA Approved: Genetically Engineered Soy and Corn in Baby Food and Nutritional Supplements* and to additional related information.

http://www.greenpeace.org/~geneng/gehome.htm

Igene Biotechnology

Igene Biotechnology focuses on the commercialization of value-added specialty biochemical products for the food, feed, flavor and agrichemical industries. Through the use of state-of-the-art fermentation technology, physical and chemical separation technology, and related chemical and biochemical engineering technologies, the company has developed proprietary processes to convert selected agricultural raw materials or feedstock into commercially useful and cost effective "environmentally friendly" natural alternatives to products now produced by chemical synthesis.

http://www.igene.com/

InfoBiotech Canada (IBC)

IBC is a partnership of government, private and academic sectors with the goal of providing enhanced access to information on Biotechnology in Canada and worldwide.

http://www.cisti.nrc.ca/ibc/home.html

Information Systems for Biotechnology (ISB)

ISB provides information resources to support the environmentally responsible use of agricultural biotechnology products. Here you will find documents and searchable databases pertaining to the development, testing and regulatory review of genetically modified plants, animals and microorganisms within the U.S. and abroad.

http://www.nbiap.vt.edu/

Insect Biotech Canada Inc. (IBC)

Insect Biotech Canada Inc. (IBC) is a Canadian-based, non-profit biotechnology research and development consortium focused on the application of biotechnology to insects and their pathogens. IBC has brought together Canada's top researchers in insect physiology, biochemistry, insect virology and molecular biology in an integrated and coordinated organization. The network is focused on serving the needs of Canada and the international community in identifying and controlling insect pests using current techniques of biotechnology.

http://www.zoo.utoronto.ca/insectbiotechcan/

Institute of Biosciences and Bioengineering, Rice University

The Institute's mission is to promote cross-disciplinary research and education encompassing the biological, chemical, and engineering disciplines. The Institute represents a unique educational environment in which to prepare students to meet the complex scientific, technological, and ethical challenges they face as they head into the twenty-first century.

http://ibb.rice.edu/

InterSciences Inc.

InterSciences is a privately-owned and operated, scientific instrumentation and supplies marketing organization representing specialized principal manufacturers lines. InterScience's primary marketing focus is Biotechnology for academic and hospital research, industrial, government, private lab and Pharmaceutical segments of the Canadian market.

http://www.interscience.com/

Leinco Technologies

Leinco Technologies is a global biosciences company with manufacturing facilities. The company is built on a diverse biotechnology platform and offers product development and contract manufacturing services under GLSP and cGMP for active biopharmaceutical intermediates. The company also offers an extensive line of antibodies, recombinant proteins and reagents for research in immunology, cell biology, and neuroscience.

http://www.leinco.com/

Link Technologies Ltd

Link Technologies Ltd, established in Glasgow in 1989, is a growing biotechnology company that now operates at the Strathclyde Business Park. The company, which has a strong reputation in the U.S. and European markets for manufacturing quality chemical reagents for the synthesis and modification of DNA and RNA, is now also developing innovative reagents for use in related areas of molecular biology, such as the structural study of proteins and the synthesis of oligonucleotide-peptide conjugates. In tandem with its expanding catalogue of products, the company offers a contract synthesis service for molecular biology reagents.

http://www.linktech.co.uk/

Louisiana Alliance for Biotechnology (LAB)

LAB is dedicated to enhancing biotechnology research and business development in the State of Louisiana. The goals set for LAB in order to accomplish this mission are to facilitate: the development of an interactive consortium of university and private sector researchers in molecular biology and biotechnology; the transfer of scientific advances into the private sector; enhancement of the university academic environment and community quality of life through economic development associated with increased private sector research and development; development and funding of research to improve and expand Louisiana's agriculture and industry; coordination of state, federal, academic and business sector efforts to improve economic conditions within the state through economic development in the biotechnology sector.

http://www.labiotech.org/

LumaCare™ Biotechnology

LumaCare™ is a scientific company developing instrumentation products in the field of biotechnology. The company has developed a specialized noncoherent light source for the field of photodynamic therapy (PDT) and for universal use with new Biotech pharmaceuticals (such as Photofrin). Photodynamic therapy is a safe, non-invasive treatment for numerous diseases, including psoriasis and other dermatological afflictions.

http://www.lumacare.com/

National Biotechnology Information Facility

The National Biotechnology Information Facility (NBIF) provides a single-point access to a vast store of widely distributed biotechnology data including approximately 7,000 links to biotechnology resources. The NBIF encourages information sharing between researchers in Academia, Industry, Government, and the Department of Defense, promotes biotechnology in Historically Black Colleges and Universities and Minority Institutions (HBCU/MIs), and provides training in biotechnology. The NBIF is also active in developing new sources and types of biotechnology databases.

http: / /www.nbif.org

National Center for Biotechnology Information (NCBI)

NCBI, a national resource for molecular biology information, creates public databases, conducts research in computational biology, develops software tools for analyzing genome data, and disseminates biomedical information, all for the better understanding of molecular processes affecting human health and disease.

http://www.ncbi.nlm.nih.gov/

Nature Biotechnology Directory

The Nature Biotechnology Directory is a global information resource listing over 9,000 organizations, product and service providers in the biotechnology industry. Entries display details about each organization, including organization name, address, telephone and fax number, e-mail and web addresses.

http://guide.nature.com/

Neugenesis Corporation

Neugenesis Corporation is a biotechnology company focused on generating better biotherapeutics cost-effectively. To this end, Neugenesis has developed technology that has substantial value in discovering, improving, and producing complex proteins, especially heteromeric proteins. The company's work with the filamentous fungus *Neurospora crassa* has resulted in significant advances in two important areas of biotechnology research, recombinant protein production and combinatorial biology.

http://www.neugenesis.com/

New Zealand Biotechnology Association

The Association was established to improve knowledge of biotechnology in New Zealand; to represent the interests of those concerned with biotechnology in New Zealand; to provide a forum for interaction among the disciplines involved in biotechnology; and to foster an environment conducive to the growth of biotechnology-based industries in New Zealand.

http://www.biotech.org.nz/

North Carolina Biotechnology Center

The Center's mission is to provide long-term economic benefit to North Carolina through support of biotechnology research, development and commercialization statewide. It works toward that mission by assisting business, education, and science. The Center is active at all points in the movement of biotechnology from ideas to marketable products. It works with academic researchers, entrepreneurs and company scientists, educators, governmental policy-makers and all other parties involved in building a thriving biotechnology industry.

http://www.ncbiotech.org/

Northern Bioidentification Service Ltd.

Northern Bioidentification Service Ltd. is a biotechnology company providing DNA analysis techniques to fish and wildlife conservation managers. The company works with animal conservation managers and researchers to protect and improve natural stocks of fish and wildlife. Through the use of high-quality DNA typing and population genetic analysis, the company aims to provide powerful and cost-effective tools used in conservation management.

http://www.mts.net/~northern/

NRC Biotechnology Research Institute

BRI is located in Montréal, Canada, and maintains advanced facilities to carry out collaborative research projects in molecular biology and biochemical engineering, including: radiocrystallography equipment, an ion spray mass spectrometer, NMR spectrometers, computer-assisted molecular design, confocal light scanning microscopy system, analytical chemistry laboratory (for the environmental sector), a 1,500 m^2 bioprocess facility with fermenters ranging in size from 3.5 L to 1,500 L, and 350 m^2 of clean rooms for separation and purification (class 10,000 and 100,000). BRI concentrates its resources on collaborative projects with Canadian industry, particularly companies working in the pharmaceutical and resource sectors.

http://www.bri.nrc.ca/

NRC Plant Biotechnology Institute

The mission of the Plant Biotechnology Institute is to perform; assist and promote strategic research in plant biotechnology, to improve and diversify Canadian industry, and to strengthen Canada's competitive position in the global economy by fostering an entrepreneurial culture. The Institute performs strategic research in key areas of plant biotechnology while maintaining a world-class research calibre; ensures industrial relevance of its research programs; nurtures innovation through efficient transfer of new biotechnologies and their products to commercialization; provides infrastructure and expertise; informs and transfers discoveries in plant biotechnology to all interested parties in Canada; and trains and develops highly qualified personnel in selected areas of plant biotechnology.

http://www.pbi.nrc.ca/

Osmetech

Osmetech is dedicated to the rapid and inexpensive detection of micro-organisms such as infectious bacteria, fungi and moulds through the application of multi-sensor array technology.

http://www.aromascan.com/

Progen Industries Limited

Progen Industries Limited is an Australian biotechnology company specialising in the manufacture and supply of molecular biology reagents and kits. Involved in pharmaceutical research and development, Progen recently commenced its second Phase I clinical trial with its lead compound (PI-88) for treatment of solid tumors. Progen is also involved in the development of a number of other potential biopharmaceuticals.

http://progen.com.au/

Toronto Biotechnology Initiative (TBI)

The TBI is a non-profit organization committed to promoting the growth of biotechnology in Toronto and the surrounding region. While biopharmaceuticals and diagnostics are core strengths of this community, it also has broad research capability and growing commercial activity in agri-food biotechnology as well as the emerging resource and environmental biotech industries.

http://www.torontobiotech.org/

University of Wisconsin Biotechnology Center

The University of Wisconsin Biotechnology Center was established to coordinate the vigorous development of biotechnology on the UW-Madison campus. As the UWBC grew, a service facilities was established to provide excellent shared resources for the UW-Madison, state and national biotechnology research communities. Outreach functions were developed, including technology transfer, biotechnology education for the public, and the gathering and dissemination of information through the Center's newsletter, *Wisconsin BioIssues*, and through the Internet. Multidisciplinary applied research programs were started (including biopulping, bioremediation, and the production of industrial enzyme in transgenic alfalfa) that were supported by federal and state grants and industrial consortia.

http://www.biotech.wisc.edu/

Bibliography

Aparicio, S. *et al.* Whole-genome shotgun assembly and analysis of the genome of *Fugu rubripes*. *Science* 297, 1301–1310 (2002).

Avery, O. T., MacLeod, C.M.& McCarty, M. Studies of the chemical nature of the substance inducing transformation of pneumococcal types. Induction of transformation by a desoxyribonucleic acid fraction isolated from *Pneumococcus* Type III. *J. Exp.Med.* 79, 137–158 (1944).

Bush, G.W., "Stem Cell Science and the Preservation of Life," *The New York Times*, August 12, 2001, p. D13.

Cohen, S. N., Chang, A.C., Boyer, H.W. & Helling, R. B.

Collins, F. & Galas, D. A new five-year plan for the US Human Genome Project. *Science* 262, 43–46 (1993).

Collins, F. S. *et al.* New goals for the US Human Genome Project: 1998–2003. *Science* 282, 682–689 (1998).

Drews, J.& Ryser, S. The role of innovation in drug development.

Druker, B. J. Imatinib alone and in combination for chronic myeloid leukemia. *Semin.Hematol.* 40, 50–8 (2003).

Fuller, B. P. *et al.* Policy forum: Ethics — privacy in genetics research. *Science* 285, 1359–1361 (1999).

Gabriel, S. B. *et al.* The structure of haplotype blocks in the human genome. *Science* 296, 2225–2229 (2002).

Gardner, M. J. *et al.* Genome sequence of the human malaria parasite *Plasmodium falciparum*. *Nature* 419, 498–511 (2002).

Gavin, A. C. *et al*. Functional organization of the yeast proteome by systematic analysis of protein complexes. *Nature* 415, 141–147 (2002).

Golub, T.R.Genomic approaches to the pathogenesis of hematologic malignancy.*Curr.Opin.Hematol.* 8, 252–261 (2001).

Guttmacher and Mark S.Guyer are at the National Human Genome Research Institute, National Institutes of Health, Bethesda, Maryland 20892, USA.

Guttmacher, A. E.& Collins, F. S. Genomic medicine — A primer.*N.Engl. J.Med.* 347, 1512–1520 (2002).

Hannon, G. J.RNA interference. *Nature* 418, 244–251 (2002).

Hilbert, D.Mathematical problems. *Bull.Am.Math. Soc.* 8, 437–479 (1902).

Hirschhorn, J.N., Lohmueller, K., Byrne, E. & Hirschhorn, K.A comprehensive review of genetic association studies. *Genet.*

Holt, R. A. *et al.* The genome sequence of the malaria mosquito *Anopheles gambiae*. *Science* 298, 129–149 (2002).

Horii, T., Nagao, Y., Tokunaga, T., and Imai, H. 20003. Serum-free culture of murine primordial germ cells and embryonic germ cells. Theriogen. 59, 1257-1264.

Horii, T., Nagao, Y., Tokunaga, T., and Imai, H. 20003. Serum-free culture of murine primordial germ cells and embryonic germ cells. Theriogen. 59, 1257-1264.

Jackson, D. A., Symons, R. H. & Berg, P. Biochemical method for inserting new genetic information into DNA of Simian Virus 40: circular SV40 DNA molecules containing lambda phage genes and the galactose operon of *Escherichia coli*. *Proc.Natl Acad. Sci.*

Kaczorowski, D.J., Patterson, E.S., Jastromb, W.E.,

Shamblott, M.J. 2002. Review Article. Glucose-responsive insulin-producing cells from stem cells. Diabetes Metab.

Kaczorowski, D.J., Patterson, E.S., Jastromb, W.E., Shamblott, M.J. 2002. Review Article. Glucose-responsive insulin-producing cells from stem cells. Diabetes Metab.

Kalka, C. et al. Transplantation of ex vivo expanded endothelial progenitor cells for therapeutic neovascularization. Proc. Natl. Acad. Sci. USA 2000; 97:3422-3427.

Kalkman EAJ, Bilgin YM, van Haren P, van Suylen R-J, Saxena PR, Schoemaker RG.Determinants of coronary reserve in rats subjected to coronary artery ligation or aortic banding. Cardiovasc Res 1996.

Kawano H, Do YS, Kawano Y, Starnes V, Barr M, Law RE, Hsueh WA. Angiotensin II has multiple profibrotic effects in human cardiac fibroblasts.

Kehat I, Kenyagin-Karsenti D, Snir M, Segev H, Amit M, Gepstein A, Livne E, Binah O, Itskovitz-Eldor J, Gepstein L. Human embryonic stem cells can differentiate into myocytes with structural and functional properties of cardiomyocytes.

Kellerman S, Moore JA, Zierhut W, Zimmer HG, Campbell J, Gerdes AM. Nuclear DNA content and nucleation patterns in rat cardiac myocytes from different models of cardiac hypertrophy. *J Mol Cell Cardiol,* 497-505 (1992).

Kennedy M, Firpo M, Choi K, Wall C, Robertson S, Kabrun N, Keller G. A common precursor for primitive erythropoiesis and definitive haematopoiesis.

Kerr, D.A., Lladó, J., Shamblott, M.J., Maragakis, N.J., Irani, D.N., Crawford, T.O, Krishnan, C., Dike, S., Gearhart, J.D., and Rothstein, J.D. 2003. Human embryonic germ cell derivatives facilitate motor recovery of rats with diffuse motor neuron injury. J. Neurosci. 23, 5131-5140.

Kerr, D.A., Lladó, J., Shamblott, M.J., Maragakis, N.J., Irani, D.N., Crawford, T.O, Krishnan, C., Dike, S., Gearhart, J.D., and Rothstein, J.D. 2003. Human embryonic germ cell derivatives facilitate motor recovery of rats with diffuse motor neuron injury. J. Neurosci. 23, 5131-5140.

Kimura, T., Suzuki, A., Fujita, Y., Yomogida, K., Lomeli, H., Asada, N., Ikeuchi, M., Nagy, A., Mak, T.W., and Nakano, T. 2003. Development and Disease. Conditional loss of PTEN leads to testicular teratoma and enhances embryonic germ cell production. Development 130, 1691-1700.

Kimura, T., Suzuki, A., Fujita, Y., Yomogida, K., Lomeli, H., Asada, N., Ikeuchi, M., Nagy, A., Mak, T.W., and Nakano, T. 2003. Development and Disease. Conditional loss of PTEN leads to testicular teratoma and enhances embryonic germ cell production. Development 130, 1691-1700.

Klug MG, Soonpaa MH, Koh GY, Field LJ (1996) Genetically selected cardiomyocytes from differentiating embryonic stem cells form stable intracardiac grafts. J Clin Invest 98:216-224.

Kocher A, Schuster M, Szabolcs M, Itescu S. Cardiomyocyte regeneration after neovascularization of ischemic myocardium by human bone marrowderived angioblasts. Nature Medicine 2002, in press.

Kocher AA, Schuster MD, Szabolcs MJ, Takuma S, Burkhoff D, Wang J, Homma S, Edwards NM, Itescu S. Neovascularization of ischemic myocardium by human bone-marrow-derived angioblasts prevents cardiomyocyte apoptosis, reduces remodeling and improves cardiac function. Nature Med.

L Field: Future therapy for cardiovascular disease. In Proceedings of the NHLBI Workshop Cell Transplantation: Future Therapy for Cardiovascular Disease? Columbia, MD; 1998.

Labastie M-C, Cortes F, Romeo P-H, Dulac C, Peault B. Molecular identity of hematopoietic precursor cells emerging in the human embryo. Blood 1998;92:3624-3635.

Levkau B, Koyama H, Raines EW, Clurman BE, Herren B, Orth K, Roberts JM, Ross R. Cleavage of p21cip1/waf1 and p27 kip1 mediates apoptosis in endothelial cells through activation of cdk2: role of a caspase cascade.

Liechty KW, MacKenzie TC, Shaaban AF, et al. Human mesenchymal stem cells engraft and demonstrate site-specific differentiation after in utero transplantation in sheep. Nat Med 2000, 6:1282-6.

Lin Y, Weisdorf DJ, Solovey A, Hebbel RP. Origins of circulating endothelial cells and endothelial outgrowth from blood.

Lynch, H. T. & de la Chapelle, A. Genomic medicine: hereditary colorectal cancer *N.Engl. J.Med.* 348, 919–932 (2003).

Lysaght, M.J., and Hazlehurst, A.L., "Private Sector Development of Stem Cell Technology and Therapeutic Cloning, " *Tissue Engineering* 9(3): 555-561 (2003).

MacLellan WR and Schneider MD. Genetic dissection of cardiac growth control pathways Annu. Rev. Physiol. 2000. 62:289-320.

Mahon NG, O'Roke C, Codd MB, et al. Hospital mortality of acute myocardial infarction in the thrombolytic era. Heart 1999;81:478-82.

Maier S, Tertilt C, Chambron N, Gerauer K, Huser N, Heidecke C-D, Pfeffer K (2001) Inhibition of natural killer cells results in acceptance of cardiac allografts in CD28-/- mice. Nature Med 7:557-562.

Makino S, Fukuda K, Miyoshi S, et al. Cardiomyocytes can be generated from marrow stromal cells in vitro. J Clin Invest 1999, 103:697-705.

Maxam, A. M. & Gilbert, W. A new method for sequencing DNA.

McCarthy PM, Rose EA, Macmanus Q, et al. Clinical experience with the Novacor ventricular assist system. J Thorac Cardiovasc Surg. 1991; 102:578-587.

McEwan PE, Gray GA, Sherry L, Webb DJ, Kenyon CJ. Differential effects of angiotensin II on cardiac cell proliferation and intramyocardial perivascular fibrosis in vivo. Circulation 1998;98:2765-2773.

Mendel, G.Versuche über Pflanzen-Hybriden.*Verhandlungen des naturforschenden Vereines, Abhandlungen, Brünn* 4, 3–47 (1866).

Migeon, B.R., Chowdry, A.K., Dunston, J.A., and McIntosh, I. 2001. Identification of *TSIX*, encoding an RNA antisense to human *XIST*, reveals differences from its murine counterpart: Implications for X inactivation. Am. J. Human Genet. 69, 951-960.

Migeon, B.R., Chowdry, A.K., Dunston, J.A., and McIntosh, I. 2001. Identification of *TSIX*, encoding an RNA antisense to human *XIST*, reveals differences from its murine counterpart: Implications for X inactivation. Am. J. Human Genet. 69, 951-960.

Milki, A.A. et al., “Incidence of monozygotic twinning with blastocyst transfer compared to cleavage-stage transfer” *Fertility and Sterility* 79(3): 503-506 (2003).

Miller, P. S. Is there a pink slip in my genes? *J.Health Care Law Policy* 3, 225–265 (2000).

National Center for Health Statistics, "Births: Final Data for 2001." National Vital Statistics Reports 51(2) (2002), available at http://www.cdc.gov/nchs/data/nvsr/ nvsr51/ nvsr51_02.pdf.

National Institutes of Health Revitalization Act of 1993, Pub. L. No. 103-43, § 121(c), 107 Stat. 122 (1993), repealing 45 C.F.R. § 46.204(d).

National Institutes of Health, *Report of the Human Embryo Research Panel*, Bethesda, MD: NIH (1994).

National Research Council.*Mapping and Sequencing the Human Genome* (National Academy Press, Washington DC, 1988).

Nelissen-Vrancken H, Debets J, Snoeckx L, Daemen M, Smits J. Time-related normalization of maximal coronary flow in isolated perfused hearts of rats with myocardial infarction. Circulation 1996;93:349-355.

Nidorf SM, Siu SC, Galambos G, Weyman AE, Picard MH. Benefit of late coronary reperfusion on ventricular morphology and function after myocardial infarction. J Am Coll Cardiol 1992;20:307-313.

Nirenberg, M.W. The genetic code: II. *Sci.Am.* 208, 80–94 (1963).

Norwitz, E. R., et al., "Implantation and the survival of early pregnancy, " *New England Journal of Medicine* 345(19): 1400-1408 (2001)

Ogawa M, Kizumoto M, Nishikawa S, Fujimoto T, Kodama H, Nishikawa SI. Blood 1999;93:1168-1177.

Onyango, P., Jiang, S., Uejima, H., Shamblott, M.J., Gearhart, J.D., Cui, H., and Feinberg, A.P. 2002. Monoallelic expression and methylation of imprinted genes in human and mouse embryonic germ cell lineages. Proc. Natl. Acad. Sci. USA 99, 10599-10604.

Onyango, P., Jiang, S., Uejima, H., Shamblott, M.J., Gearhart, J.D., Cui, H., and Feinberg, A.P. 2002. Monoallelic expression and methylation of imprinted genes in human and mouse embryonic germ cell lineages. Proc. Natl. Acad. Sci. USA 99, 10599-10604.

Oz MC, Argenziano M, Catanese KA, et al. Bridge experience with long-term implantable left ventricular assist devices. Are they an alternative to transplantation? Circulation. 1997; 95:1844-1852.

Park, T.S., Hong, Y.H., Kwon, S.C., Lim, J.M., and Han, J.Y. 2003. Birth of germline chimeras by transfer of chicken embryonic germ (EG) cells into recipient embryos.

Park, T.S., Hong, Y.H., Kwon, S.C., Lim, J.M., and Han, J.Y. 2003. Birth of germline chimeras by transfer of chicken embryonic germ (EG) cells into recipient embryos.

Pfeffer JM, Pfeffer MA, Fletcher PJ, Braunwald E. Progressive ventricular remodeling in rat with myocardial infarction. Am J Physiol 1991.

Pfeffer MA, Braunwald E, Moye LA, et al. Effect of captopril on mortality and morbidity in patients with left ventricular dysfunction after myocardial infarction. Results of the survival and ventricular enlargement trial. The SAVE investigators. N Engl J Med 1992;327:669-677.

Pitt B, Segal R, Martinez FA, et al. Randomised trial of losartan versus captopril in patients over 65 with heart failure (Evaluation of Losartan in the Elderly Study, ELITE). Lancet 1997;349:747-752.

Pittenger MF, Mackay AM, Beck SC, et al. Multilineage potential of adult human mesenchymal stem cells. Science 1999, 284:143-147.

Pouzet B, Ghostine S, Vilquin J-T, Garcin I, Scorsin M, Hagege AA, Duboc D, Schwartz K, Menasche P (2001)

Is skeletal myoblast transplantation clinically relevant in the era of angiotensin-converting enzyme inhibitors? Circulation 104:I223-228.

Rafii S, Shapiro F, Rimarachin J, Nachman R, Ferris B, Weksler B, Moore AS, Asch AS. Isolation and characterization of human bone marrow microvascular endothelial cells: hematopoietic progenitor cell adhesion.

Ravichandran LV and Puvanakrishnan R. In vivo labeling studies on the biosynthesis and degradation of collagen in experimental myocardial myocardial infarction. Biochem Intl 1991;24:405-414.

Reich, D. E. & Lander, E. S. On the allelic spectrum of human disease. *Trends Genet.* 17, 502–510 (2001).

Rothenberg, K. *et al.* Genetic information and the workplace: Legislative approaches and policy challenges. *Science* 275, 1755–1757 (1997).

Sachidanandam, R. *et al.* A map of human genome sequence variation containing 1.42 million single nucleotide polymorphisms. *Nature* 409, 928–933 (2001).

Sanger, F.& Coulson, A. R. A rapid method for determining sequences in DNA by primed synthesis with DNA polymerase.

Sapienza, C. 2002. Commentary. Imprinted gene expression, transplantation medicine, and the "other" human embryonic stem cell. Proc. Natl. Acad. Sci. USA 99, 10243-10245.

Sapienza, C. 2002. Commentary. Imprinted gene expression, transplantation medicine, and the "other" human embryonic stem cell. Proc. Natl. Acad. Sci. USA 99, 10243-10245.

Selkoe, D. J.Alzheimer's disease: genes, proteins, and therapy. *Physiol. Rev.* 81, 741–66 (2001).

Shamblott, M.J., et al., "Derivation of pluripotent stem cells from cultured human primordial germ cells" *Proceedings of the National Academy of Sciences*. 95(23): 13726-13731 (1998).

Shi Q, Rafii S, Wu MH-D, et al. Evidence for circulating bone marrow-derived endothelial cells. Blood 1998;92:362-367.

Sidow, A. Sequence first.Ask questions later. *Cell* 111, 13–16 (2002).

Smith, L. M. *et al.* Fluorescence detection in automated DNAsequence analysis. *Nature* 321, 674–679 (1986).

Smythe GM, Grounds MD (2000) Exposure to tissue culture conditions can adversely affect myoblast behaviour in vivo in whole muscle grafts: implications for myoblast transfer therapy. Cell Transplant 9: 379-393.

Smythe G.M., Hodgetts S.I., Grounds M.D. (2001) Problems and solutions in myoblast transfer therapy. J Cell Mol Med 5:33-47.

Stockwell, B. R. Chemical genetics: Ligand-based discovery of gene function. *Nature Rev. Genet.* 1, 116–125 (2000).

Takahashi, T. et al. Ischemia- and cytokine-induced mobilization of bone marrowderived endothelial progenitor cells for neovascularization. Nat. Med. 1999; 5:434-438.

The Chipping Forecast II. *Nature Genet.* 32, 461–552 (2002).

The Mouse Genome Sequencing Consortium. Initial sequencing and comparative analysis of the mouse genome. *Nature* 420, 520–562 (2002).

Tyson, J.J., Chen, K. & Novak, B.Network dynamics and cell physiology.*Nature Rev.Mol. Cell Biol.* 2, 908–916 (2001).

Van der Weyden, L., Adams, D. J. & Bradley, A. Tools for

targeted manipulation of the mouse genome. *Physiol.Genomics* 11, 133–164 (2002).

Wagner, K.R. Genetic diseases of muscle.*Neurol. Clin.* 20, 645–678 (2002).

Watson, J.D. & Crick, F.H. C. Molecular structure of nucleic acids: A structure for deoxyribose nucleic acid. *Nature* 171, 737 (1953).

Wilcox, A.J., et al., "Incidence of early loss of pregnancy," *New England Journal of Medicine* 319 (4): 189-194 (1988).

Xu, R.H., et al., "BMP4 initiates human embryonic stem cell differentiation to trophoblast" *Nature Biotechnology* 20(12): 1261-1264 (2002).

Zhang, M.Q.Computational prediction of eukaryotic proteincoding genes. *Nature Rev. Genet.* 3, 698–709 (2002).

Index